Proposal Writing Across the Disciplines

Brian R. Holloway
Mountain State University

Upper Saddle River, New Jersey
Columbus, Ohio

Editor in Chief: Stephen Helba
Executive Editor: Frank I. Mortimer, Jr.
Editorial Assistant: Barbara Rosenberg
Production Editor: Louise N. Sette
Production Supervision: Custom Editorial Productions, Inc.
Design Coordinator: Diane Ernsberger
Cover Designer: Linda Sorrells-Smith
Cover art/photo: Corbis Stock Market
Production Manager: Brian Fox
Marketing Manager: Tim Peyton

This book was set in Gill Sans by Custom Editorial Productions, Inc. It was printed and bound by R. R. Donnelley & Sons Company. The cover was printed by The Lehigh Press, Inc.

Pearson Education Ltd.
Pearson Education Australia Pty. Limited
Pearson Education Singapore Pte. Ltd.
Pearson Education North Asia Ltd.
Pearson Education Canada, Ltd.
Pearson Educación de Mexico, S.A. de C.V.
Pearson Education—Japan
Pearson Education Malaysia Pte. Ltd.
Pearson Education, *Upper Saddle River, New Jersey*

Copyright © 2003 by Pearson Education, Inc., Upper Saddle River, New Jersey 07458. All rights reserved. Printed in the United States of America. This publication is protected by copyright and permission should be obtained from the publisher prior to any prohibited reproduction, storage in a retrieval system, or transmission in any form or by any means, electronic, mechanical, photocopying, recording, or likewise. For information regarding permission(s), write to: Rights and Permissions Department.

10 9 8 7 6 5 4 3 2 1
ISBN: 0-13-022495-2

Preface

The researched writing taught in classrooms unites with the technical communication of the workplace to produce a special mode of presentation, the proposal. This mode appears in a variety of situations. For example, a student seeks formal approval of a research project. A software firm attempts to sell its network management product. A local PTA chapter, wanting equipment for its school, corresponds with the gifts-in-kind department of a large corporation. A scientist applies for a major grant. An architectural firm establishes that a structure can be built and forwards that information to its client. These tasks all require writing documents that certainly are different, but that share common characteristics and a philosophy of construction: the welding together of skills learned in business and the liberal arts.

Such writing demonstrates that the goals of a project are achievable, and usually requests specific support for that project. That is, this communication informs and persuades. Short or long, technical or plain-spoken, proposals are a vital component of workplace and institutional discourse. Understanding the patterns and the formal requirements used to create proposals will benefit you in your personal life and in your career, whether you are in sales, research, a civic organization, college, or a company exploring opportunities for growth. And, of course, our eclectic age requires that we learn about the kinds of proposal writing beyond our immediate specializations.

While teaching and using such technical communication, I've noticed few compact, general guides to proposal writing. Yet two things are likely. First, people will probably change careers several times. Second, many of those careers—particularly sales, management, science, teaching, public service, social services, or healthcare—will involve writing proposals.

I wanted a workplace-friendly and student-friendly book that could be studied at the task, whether writing a short prospectus or a complex, multipart report. When I asked colleagues in business, humanities, technology or the health services about this idea, they, too, liked the concept. Some of their documents appear in this book.

I also wanted readers to meet the creators of certain proposals, to learn about the writers' intentions, challenges, and goals—in a sense, to understand the writers as real people producing real communication with practical effects. So this book includes a few interviews of contributors as well as commentary on their work.

Though no guidebook can provide exhaustive description and discussion of all possible variations, *Proposal Writing Across the Disciplines* exhibits the techniques of a special craft in which the practical embraces the theoretical to achieve results.

The plan of each chapter is straightforward. First you will see a group of questions about the task to be discussed. Next appear the documents that answer the requirements of the task, followed by advice about how to construct these proposals. After a set of questions for further study, a review of proposal patterns closes the chapter.

You do not have to be a student in a technical writing class to use this book. The models and discussion will help you to write a proposal whether you are taking a course or are studying on your own.

May this text guide you on your way to success!

Acknowledgments

This guidebook would not have been possible without the energizing input from my students who inspired this text and for whom this book is written. Thanks to Mountain State University, too, for its encouragement.

The professionalism and insight expressed by those in different disciplines as they pursue their goals have been major influences. Thanks especially to Douglas M. Burns, Neil Manning, and Kevin Nash for their ideas. Steve Helba, editor in chief, Frank Mortimer, executive editor, and the staff at Prentice Hall have believed in this project and provided professional guidance and support, for which I'm grateful.

Many people graciously contributed material for this book—interviews as well as documents to illustrate techniques. I regret being unable to use everything. I thank Bud Branch, Amanda Richmond, Sam Deem, Emily Montgomery, Teresa Massie-Workman, Kim Weatherly, Aimee Salmons, Rachel Lanier, Paula Fields, Janice Clifford Wittekind, Neil Manning, and Doug Burns for their work appearing in this text.

I would also like to acknowledge the reviewers of this text: Helene M. Lamarre, DeVry Institute of Technology—DuPage; Douglas Gray, Otterbein College; Kenneth L. Mitchell, Southeastern Louisiana University; William J. Leonhirth, University of North Florida; Janet H. Carr, Northeastern University; and Maggie Joabar, Bays Mill Community College.

As always, my family deserves special thanks for its patience with my projects.

Contents

1

Lessons from Business: Basic Types of Proposal Writing

Business communication provides the format, strategy of presentation, and organizational underpinning of proposal writing. In this section you will learn about the basics of proposals while studying an example of common business forms—the letter, memo, and report. Memo and report styles in particular influence the long proposal, to be discussed later in this book. And these styles have applications across the disciplines, not just in business.

Questions

◆ How do I write short letters and memos proposing a plan or project?
◆ How might I extend the size of such a proposal so that it becomes a larger document with more detail?
◆ What related forms are there?

Applications

Proposals and Related Documents

Proposals Exist on a Continuum. At one end of the continuum is the verbal "pitch" of an idea to a co-worker, supervisor, or instructor. If the other party approves the suggestion, the proposal has succeeded. However, the more complicated the concept, the greater the likelihood that its champion must commit it to paper, or an electronic equivalent, so that all may understand. In some cases a brief e-mail suffices. In others, a letter or memo will do.

Longer documents, though, are the effective tools for marshaling facts, accounting for readers' needs, presenting organized concepts, and demonstrating fluency in the subjects discussed. Such documents, of course, can be e-mail attachments or electronic reports supported by multimedia as well as conventional "white papers." While we must allow for differences within variant forms, the philosophy of constructing proposals depends on similar principles of document construction.

Proposals and Feasibility Reports. If the term "proposal" includes all of the communications described above, it also embraces documents varying in purpose from the one seeking to establish whether or not a course of action is viable—a feasibility report—to the sales presentation using traditional persuasive techniques.

Example 1. A Proposal in Letter Form

What It Looks Like. You may send a letter proposing a project to someone familiar with your work, in which case the contents of the letter would reflect your client's familiarity with what you have done. For example, you might refer to your other projects, their successful completion, and their similarity to the current task being considered. Or, a proposal might be a "cold" solicitation in which you must establish credibility with the reader in order to be accepted. In either event, focus clearly on the job to be accomplished. In an opening paragraph, the letter should state your mission. Following paragraphs should detail your proposal and establish your credibility. A closing paragraph must "ask for the sale" by requesting acceptance of your project and explaining that you look forward to a reply. Be sure to make it easy for the recipient to reply to you by providing any contact information necessary. Figure 1.1 illustrates a standard letter format.

How You Do It. Select a letter format—block, semi-block, or simplified business style. (Refer to the Appendix for information about constructing business letters and adhering to their requirements.) Remember that many people consider block style to reflect a businesslike directness, whereas semi-block construction with its indented beginnings of paragraphs and its inset elements may provide a vehicle of communication that approaches the personal letter in format. The effect to be required depends on your goal and your audience. Next, use the opening paragraph to state the essentials that you wish to establish without entangling the reader in distracting detail. Subsequent paragraphs in the body, explaining the reasons for the project, should establish the necessary specifics. The concluding paragraph should request a decision forthrightly, without the presence of language that seems to beg or plead. Choose a closing element carefully: you might want to end "cordially" or "sincerely" depending on your audience and how well it knows you.

Figure 1.1 Proposal in Letter Form

DR. RAY R. VERMILLION
Department of Fine Arts

The College of Expansion
55 West Oak St.
Airdale, MO 62580

Telephone 555-555-5555—Ext. 5555; E-mail vermill@expan.edu

October 8, 2002

Lon Holloman
Editor
Photo Press
555 Caine Avenue
Wasson, OH 45555

Dear Mr. Holloman:

After talking yesterday with Klein Nathan, an executive at Photo Associates, Inc., who has seen *Photographic Basics* and liked its presentation of black and white photography, I decided to send you a text proposal.

Mr. Nathan states, and I agree, that the academic and business world might benefit from a similar book entirely devoted to the art and craft of digital photography. Such a book should be accompanied by a portfolio containing the templates of examples in different fields, so that students and those in the workplace can reference these models.

Last summer, I surveyed several instructors of photography concerning such a concept. They thought I should start on the project right away, and agreed to contribute material to the book. These teachers, all working in community colleges in Missouri, believe this effort has merit.

Please let me know what you think of this idea. I can be reached at the address above, and will send a detailed prospectus of the text upon request.

Sincerely,

Ray R. Vermillion

Ray R. Vermillion

encl.

Example 2. A Short Investigative Proposal Written as a Memo

What It Looks Like.   Sometimes a short proposal as a memo to a supervisor or colleague appears as a result of preliminary investigation. In the example shown in Figure 1.2, an employee has written such a memo to her manager about a problem that she has noticed. She has already uncovered some information as a result of her inquiries, but the bulk of her findings will appear in a formal report that she requests permission to write.

Figure 1.2   Investigative Proposal

MEMO

To:	Art Smith, Human Services
From:	Karen Schuster *KS*
Date:	January 28, 2000
Re:	Proposed Researching of Day-Care for Our Workplace

Focus. Our company is experiencing a surge in employment, but the new hires in entry-level jobs are increasingly older. Many now are parents with obligations—either

- Single mothers restarting in the workforce after an absence (average age, 25), or
- Couples beginning different careers after corporate downsizing (average age, 32).

These parents experience a major problem when our company does not answer a basic need: What to do with their young children? I propose researching the feasibility of on-site daycare for the children of our employees.

Description of Investigation. I will assess the potential for establishing a day-care facility at our location. By telephone, I will survey companies with a size and employee base similar to ours to see how they manage child care issues. I will use the Internet and any available databases to determine how other firms address this challenge. Having selected companies that appear to have solved this problem successfully, I will visit them to study their solutions, to be incorporated into a proposed model for our workplace. I will forward my recommendations to you within one month of your approval.

Approval. Please let me know if you approve the project so that I may adjust my schedule accordingly.

How You Do It.   Start with an outline similar in approach to that of the proposal letter above. Format this memo by using as headings the phrases in the outline that produced it. These should be parallel expressions. In this example, the headings are Focus, Description of Investigation, and Approval. These were also the key words in the document's outline. They provide the reader with a clear understanding of what will be covered in the communication. Begin with a message explaining the focus of the project. Use the "Focus" paragraph to establish and illustrate the problem to be studied. To strengthen your argument, use important facts from the information you have gathered. That will get the attention of the reader. Next, describe the project in the body of the memo. Finally, request approval to continue the project. Follow an outline that looks like this:

◆ Introduce the research question or challenge.
◆ Describe the proposed project.
◆ Request approval.

This pattern within its memo format can be extended to become a proposal that provides greater detail than a simpler presentation can contain. As such writing increases in size, the basic three-part structure of written communication subdivides, creating additional segments within the document in order to clarify items for the reader. Perhaps you do not merely need an introduction-body-conclusion approach, but must have a section defining key terms as well. That section will branch off from and usually follow the introduction. Perhaps your introduction itself will subdivide to address the audience with a question, illustrate the point, and state the mission of the proposal. The sections of the body themselves may become more individualized to present more complex information. Frequently, those sections are headed by questions about the specifics of the subject discussed, as shown in the next example.

Example 3. A Feasibility Study in Report Form

What It Looks Like.   The example shown is a medium-size investigative report, or feasibility study. The purpose of this proposal is to establish that something is or is not achievable, and to communicate that information to the decision-maker within the organization. In the beginning days of ISO 9000, an international quality-control standard, many companies investigated whether compliance with ISO would harm or help them. For some, the extra expense was acceptable, because of participation in international markets using this standard. For others not concerned with accessing such markets, expenses outweighed benefits. Decision-makers relied on the research of the economic environment and of their competition to determine whether their companies should participate in the ISO certification program. Written for such a decision-maker, a short proposal to retain the status quo after examining the feasibility of adopting ISO standards might have used the approach of Figure 1.3.

Figure 1.3   ISO 9000 Report by Bud Branch, 1994

TO: Schmendrick Electronics Management
FROM: Bud Branch, Quality Assurance Group
DATE: December 9, 1994
RE:

ISO 9000

Summary of Report

This report explains the features of ISO 9000, surveys the positive and negative impacts of such certification, and recommends further analysis before our company adopts this potentially costly registration.

What Is ISO 9000?
In an increasingly global economy, the European market has become concerned about the goods and services available to it. Many companies—and countries—have had their own sets of standards for their respective products. In 1987 the European group International Organization of Standards (ISO) developed a set of guidelines for building a quality system adopted by the European Economic Committee; thus, the inauguration of ISO 9000 to warrant uniformity among products produced and marketed internationally.

What Is Its Goal?
Experts believe that such uniform standards will generate security about the reliability of articles produced under their guidance. The burden of final quality control will be lifted from the recipient of such goods (Jordan 547).

Instead of having to meet several different audits and as many different quality standards, ISO 9000's goal is to have the world market comply with one set of global standards. Customers will know what they are getting when purchasing a product and/or service from an ISO 9000 certified company. This certification will be maintained by third-party organizations that have been certified by another third-party organization.

In addition, certified companies should incur increased efficiency in their operation through the reduction of waste and scrapped material. Also, improved product reliability should prevent customers from having to purchase as many spare parts and forestall excessive downtime.

Figure 1.3 *(continued)*

The Three Types of Certification

ISO 9000 and 9004 are not the designations of certifications, but provide the principles for management of an organization in the development of its quality control manuals. It is through these manuals that the company will seek and maintain the following certification in one of the three areas listed. Diversified corporations will require different certification levels based on the product and/or service offered by different sites.

- ISO 9001 is the most comprehensive certification covering everything the company designs, develops, produces, and installs.
- ISO 9002 covers development, production, and installation only.
- ISO 9003 covers only final product testing. It does not include any procedures prior to the final testing before shipping. Because few companies obtained ISO 9003 certification in 1993, this certification may be discontinued.

What Is Life Like After ISO 9000 Certification?

One fear is the impact this program could have on a company and its operation. Auditing and assessment may take precedence over pleasing customers (Jordan 547).

Also, small companies with innovative designs may not be able to develop the procedures and follow-up required by ISO 9000.

Although pricing should not be affected in principle, the cost of registration is expensive—involving the development of procedures, documenting them, and auditing them. Once these systems are set up, savings are supposed to be gained through increased efficiency (fewer rejects, reworks, inspections). Although more customers and repeat business may offset the initial expenses, the customer should not necessarily expect price reductions as a result.

Many ISO certified companies surveyed for this report privately state that the systems they have developed in order to obtain certification are not fully in place after the registration audit. These organizations feel that at least two subsequent audits are needed before their documentation can be refined and understood by employees. Certification, then, becomes a process, not a prize awarded for a one-time inspection: the registrars come back every six months to follow up. The participants must understand the spirit of the standard in order to comply.

Because plans include product reliability, availability, and maintainability, the producer must furnish documentation ensuring these factors. But some companies state that the amount of documentation they must create stifles their

---

(continued on next page)

Figure 1.3   *(continued)*

flexibility, creates more work, and erects barricades to change. When a company overdocuments, it also must live with the inconvenient results, since what is declared in the paperwork must be accomplished (Mullin and Kiesche 34-40).

Conclusion

The articles and companies surveyed for this report affirm that if an organization obtains ISO 9000 registration and then loses it because of noncompliance, such would be worse for that company's image than being unregistered in the first place. Critics of ISO 9000 also argue that because it requires documentation only of procedures, it may ensure that a company with a mediocre level of quality will stagnate at that level. Of course, proponents believe recertification should be a continuing process of improvement of any company's operation and product. Yet many questions arise about the cost of certification and the reality of any potential payback.

Recommendation

Because of the program's newness, one cannot really perceive its ramifications. Our main competitor is not ISO certified, and does not directly compete with us in the same global markets. I recommend further research and much caution before we take the expensive step toward certification.

Print Sources Surveyed for this Report

"After ISO 9000." *Electronic Business Buyer* October 1993: 48-64.

Jordan, Jo Rita, "ISO 9000 and Analytical Instruments." *LG-GC* August 1993: 547.

Mullin, Rick, and Elizabeth S. Kiesche. "ISO 9000: Beyond Registration." *Chemicalweek* April 20, 1993: 34-40.

Rabbitt, John T., and Peter A. Bergh. *The ISO 9000 Book: A Global Competitor's Guide to Compliance and Certification.* White Plains, NY: Quality, 1993.

Sanders and Associates. *How to Qualify for ISO 9000.* New York: Sanders/American Management, 1993.

How You Do It. Outline the main points to cover by listing the key phrases governing your discussion. You will find that questions are often best to use in this sort of outline, because they force the writer to respond to them in clear paragraphs. You will then use these outline phrases as the headers for sections in your proposal. You will need a summary at the beginning that will tell the reader the key recommendations of the proposal. Follow it with definitions of ISO standards. Explain their goals. Describe the standards and their implications in detail. Wrap up the discussion and state your recommendations. Close with a list of references consulted in producing the report. Use a plan to guide your discussion that follows this pattern:

◆ Introduce the gist of the study.
◆ Define ISO standards.
◆ Review ISO goals.
◆ Survey the implications of ISO using examples from comparable firms.
◆ Review the company's internal readiness to comply with ISO.
◆ Wrap up by summarizing your findings.
◆ Recommend deferring complying with ISO requirements.
◆ List the references used in compiling the report.

Advice

Don't Worry If the Form Is New to You. Many people who find themselves writing proposals have not originally been trained to produce them. Nor are they necessarily working in a corporate environment. However, for support personnel, managers, professionals, and college students, the need to construct such documents may suddenly arise. Remember that such writing is not that different in purpose from much of what we do daily, informally, when we want to accomplish something. We constantly suggest ideas, support them with data, and urge actions, as do written proposals. Furthermore, proposal writing is governed by simple patterns within a style that you can learn to follow. These patterns, which we begin exploring here, will appear throughout this book. They must be clear because the documents governed by them are read by busy people who must understand quickly.

Know Your Audience and Its Protocol. Crucial in the success of such documents is an understanding of audience and audience expectations. Proposals belong to the realm of business and technical writing. Because they are parts of this content area, they must fulfill the standards of a "discourse community," a group of people sharing the same means of and notions about communication. Not knowing the rules of the community could penalize an author who might commit infractions detracting from a presentation. For example, submitting handwritten grant proposals, like submitting handwritten cover letters applying for jobs, is such a violation of the rules. Proposals follow styles accepted by companies or institutions, and employ the language of the particular field under discussion.

Understand the Way Readers Conceptualize.   Some learning theorists have speculated that there are different types of intelligence. Certainly, everyone learns and forms concepts differently. Some can be told about an idea and then understand it. Some have to be shown. Some require a strong visual component because they think in pictures. To appeal to your audience's diverse characteristics of reception, it is best to construct proposals so that they tell readers about the issue, show readers examples supporting the points made, and employ pleasing layout and graphics to enhance the presentation.

Consider Persuasive Needs.   Even if a proposal reflects an investigation of data to show whether something is feasible, it is persuading—by summarizing, quoting, and analyzing research to present the case that the project is worthy of consideration or that it can be accomplished. Your central questions in designing a proposal are: What do I want my reader to do? What does my reader need to find out in order to act? Generally the investigation reflected in proposal writing is assisted by standard strategies of persuasion.

Understand Persuasive Techniques.   Ultimately, any writing is persuasive, but proposals often explicitly employ persuasive techniques first presented in antiquity. Figure 1.4 displays the ancient and current uses of these concepts. It is important to note that in keeping with these approaches a proposal may:

◆ begin with a summary establishing the initial relationship with the audience, getting its attention. This introduction may then
◆ illustrate the point of the proposal, perhaps declaring key concerns, and then will
◆ state the document's purpose. The body of the proposal may then
◆ partition the areas to be discussed, then discuss them. The closing sections then
◆ provide the recommendations of the proposal and may request approval.

These functions are all implicitly present in very short proposals, but appear independently and often in this sequence in larger documents.

Use Resources.   Although writers may learn profitably from their large-scale mistakes, the proposal represents too significant an investment of labor for this to be an ideal educational method. Reviewing what is expected in business and technical writing will help prevent problems as you develop your documents. Often your firm or institution has a published "house style" that you can follow—and if it does not, it should. Your dictionary and handbook of grammar and usage will help. The Appendix to this book contains a survey of the style, audience, tone, structure, format, and documentation characteristic of business and technical writing.

Figure 1.4 Structures of Persuasion

Persuasive presentations (essays, speeches, reports) borrow their organization from classical rhetoric. In classical rhetoric, a presentation has the following parts:

> **Exordium.** The first part of the introduction is an *exordium,* or a direct acknowledgment of your audience.

> **Narratio.** Then follows an illustration of the subject, which leads to—

> **Propositio.** The thesis of the presentation. This ends the introduction.

> **Divisio.** The first part of the body defines key terms, or presents the structure of the argument to the audience ("divides" the discussion).

> **Confirmatio** and **Confutatio.** These parts of the body explore problem/solution, or pro/con. This is the heart of the argument.

> **Conclusio.** The conclusion, which should refer to the **narratio** if possible.

In modern persuasive writing, we can use these ancient divisions of presentation to control our argument:

> **Introduction.** A good introduction does three things—it
> *Addresses* the audience
> *Illustrates* the problem
> Proposes the **thesis.**

> **Body.** The body of a problem/solution paper
> Begins with a section that **defines** unfamiliar terms or shows the **division** of the argument to come
> Discusses the **problem**
> ***Discusses and evaluates the solutions.***

> **Conclusion.** The conclusion calls for action or belief and—if artful—refers back to the **illustration** used in the introduction.

For Further Study

1. What academic or community situations could involve the use of proposals? Consider such things as a college project, a PTA enhancement for an elementary school, a proposition to improve community safety brought before a town council, or the organizing of a volunteer effort.

2. What format of proposal (letter, memo, or report) would be best for which kinds of tasks? For example, would a life insurance agent first contact a customer by sending a report?

3. For what purposes could some of these formats combine to work together?

4. How might you combine them?

5. Any of these three types of short proposal can be sent as an e-mail attachment and retain an appearance identical to that of its printed paper copy. Also, because e-mail templates are similar to the memo in form, a short informal proposal can appear as an e-mail itself. Write one. Consider audience and strategy in so doing, and study Figure 1.5.

Figure 1.5   A Short Proposal by E-mail

Subject:	**Day-Care Project**
Date:	Fri, 13 Jul 2000 07:23:16 -0500
From:	Art Smith, Human Services
To:	Sharon Neville, VP

Sharon—

I propose we proceed with Karen Schuster's recommendation for Phase One: to do the light remodeling necessary to convert designated floor space in BLDG 1, and to bring in Fritillary Daycare to operate the facility part-time.

Using Fritillary, an independent contractor with experience, will permit us to concentrate on the main functions of our company while providing a needed, though basic, beginning of day-care coverage for our employees' children.

I recommend a thorough, ongoing evaluation of the contractor's services while we discuss the proposed Phase Two, in which we integrate all functions of day-care into our company.

You should receive today via interoffice mail a second packet of information that Karen Schuster developed. This contains raw data, but will assist us as we plan Phase Two.

Please let me know if these ideas meet with your approval.

—Art

Review of Patterns

Letter

Company or sender's address

Date

Recipient's address

Salutation

Opening paragraph stating the essential message

Subsequent paragraphs in the body

> Explaining the reasons for the project

> Establishing the specifics necessary

Concluding paragraph requesting a decision

Closing line

Signature

Typed name of sender

Memo

The heading "Memo" or "Memorandum"

Beginning elements

> To:

> From:

> Date:

> Re:

Paragraph stating the message

Paragraphs supporting the message and describing the proposed project

Paragraph requesting approval

Short Report

Introduction stating mission

Definition of key terms

Review of project goals and methods

Analysis of positive and negative impacts of proposed project

Concluding summary

Recommendations

List of references used

E-mail

Information about sender and recipient

Message statement

Discussion of reasons for the proposed plan

Conclusion with recommendations

2

Proposals on Campus

This chapter examines short proposals used in academic settings. Borrowing format and style from business presentation, the proposal, or prospectus, is an integral component of many courses and projects.

Questions

◆ How do I target the proper audience for my proposal?
◆ How do I display content in the proposal?
◆ What formats do I use?
◆ What decides this?
◆ Does the proposal template appear in other types of academic writing?

College is the place where many first encounter proposal writing, and such writing can be challenging to those trained to produce a different type of document, such as a paper for an English class. For example, in an English essay, the outline is concealed. The contents flow according to the plan you have devised, and transitions move the reader from one section to another, but the reader does not usually view the outline itself while surveying the text of the paper. However, a proposal is both an outline and a unified collection of short presentations that each respond to a point in the outline. A proposal displays— even accentuates—its template, the parts of which are used as headers to control the flow of discussion. It could be said that an English essay has an endoskeleton and a proposal has an exoskeleton.

Applications

Proposals appear in the classroom when a student drafts a prospectus for a research project. Another frequent use of the proposal form is as a response to requests for conference presentations or related activities. In addition, proposal format may help organize the content of a long research paper with an agenda that features a call to action. This feature is intuitively familiar to every undergraduate student who has written a problem/solution essay. Moreover, the concepts of proposal writing may shape a request to proceed with an extended academic project. The proposal can thus appear as a prospectus for an extended research program—a project occupying several years of the student's or professor's time, in the case of a thesis proposal, or one inaugurating a long, grant-funded period of study. Since obtaining grant funding is itself a business, the documents proposing funded projects can be quite businesslike themselves—something we will explore later in this text. In this chapter, we will look at examples of academic documents that depend on the proposal pattern. And—because context is important—we'll discover the people who created some of these proposals, learn how the proposals were developed, and consider the implications of this communication.

Example 1. A Short Proposal for a Research Project

What It Looks Like.   Figure 2.1 presents a document generated to obtain approval for a term research project in a college course. The example demonstrates proposal writing in a short memo form. Note that the parts of the proposal can be used as the basis for an outline of the presentation itself.

How She Does It.   Amanda is using an expanded memo form to create this research prospectus. She replaces the word "Memo" with the term "Research Proposal," however, so that the reader can immediately identify the purpose of the document. In addition, the subject line of the header identifies the content of her project. One could, of course, title the proposal differently, replacing "Research Proposal" with the title of the project. Then, the subject line in the header could be eliminated, as suggested by Figure 2.2 on page 19.

The plan of Amanda's prospectus ensures a pleasing presentation on the page. Because such visual relief is important in convincing the reader, the document contains several elements that promote eye movement and vary the appearance of text. These include

◆ Major and subordinate headings in boldface type, which alert the reader to important ideas and show the relative hierarchy of such concepts
◆ Indenting of subordinate material, which cues the reader to the presence of such material and lets the reader's eyes shift diagonally to the right when skimming the document
◆ Inserting an extra line between major sections of the report, which separates concepts visually, conveying the impression that the paper is organized and easy to digest
◆ Choosing a font that presents a clean, clear appearance, free from distracting curlicues and flourishes.

Figure 2.1   Proposal by Amanda Richmond

RESEARCH PROPOSAL

DATE: March 30, 1999
TO: Dr. Brian Holloway
FROM: Amanda Richmond
SUBJECT: Fresh Water Pollution

INTRODUCTORY SUMMARY

I plan to write my report on fresh water pollution. My goal is to provide a
detailed look at the effects of fresh water pollution, stressing the damage
caused by agricultural runoff. I will discuss the causes and effects of fresh
water pollution, provide possible solutions, and give examples of how this
type of pollution is being treated.

AREAS TO BE STUDIED

My research will contain information on the following topics:

- **Why is fresh water important?**
  In the first section, I will reveal the importance of each fresh water source.
  I will also explain how fresh water sources benefit human life and then dis-
  cuss the effect pollution has on the source and environment.
- **What causes fresh water pollution?**
  In this section, I will discuss the various types of fresh water pollution.
  Then, I will focus on the nation's number one source of fresh water pollu-
  tion, non-point source pollution from agricultural runoff.
- **What are the effects of fresh water pollution?**
  This section will identify several bacteria, viruses, and pathogens that
  pollute fresh water and reveal the effects they have on humans and the
  environment.
- **How can pollution from agricultural runoff be cleaned up and
  prevented?**
  Finally, I will describe the process of how to identify the source of pollution. Also, I will explain how to implement a plan to clean up polluted water
  and prevent future pollution. Examples of different water management
  plans that are being used to control pollution in surrounding areas will also
  appear in this section of my report.

METHODS OF RESEARCH

First, I went to the CWV Learning Resource Center and found several books
using Athena. The books contained general information on the effects of fresh
water pollution. Then I searched the EBSCOhost database and located magazine articles on the importance of fresh water and where it comes from.

(continued on next page)

Figure 2.1 *(continued)*

The Newsbank database provided me with numerous newspaper articles on local pollution problems and methods of clean up that are being used in surr- ounding areas. Yet the best source of information on non-point source, fresh water pollution occurring from agricultural runoff came from the Soil Conservation District. This information was in the form of pamphlets, maga- zines, educational packets, and specific case studies on fresh water pollution.

My sources thus far are:

Cairns, John, et al. <u>Fresh Water Pollution III: Problems & Control</u>. New York; MSS Information Corporation.

Chanlett, Emil T. <u>Environmental Protection</u>. New York; McGraw-Hill, 1973.

D'Itri, Frank M. and Louis G. Wolfson. <u>Rural Groundwater Contamination</u>. Chelsea; Lewis Publishers, Inc., 1988.

Hendricks, Charles W., et al. <u>Fresh Water Pollution I: Bacteriological and Chemical Pollutants</u>. New York; MSS Information Corporation, 1973.

Knowles, Forrest E., et al. <u>Fresh Water Pollution II: Radioactive Pollutants</u>. New York; MSS Information Corporation, 1973.

Krautz, Joachim. "Poisoning the Fount of Life–Fresh Water Pollution and its Consequences." <u>Contemporary Review</u> 265 (Sep 1994): 144–48.

Leavenworth, Stuart and James E Shiffer. "Airborne Menace." <u>News and Observer</u> 5 July 1998, final ed.: A1.

McCall, Julian and Janice Crossland. <u>Water Pollution</u>. New York; Harcourt Brace Jovanovich, Inc., 1974.

Melnykovych, Andrew. "River on the Rebound—Runoff from Farmland Holds a Host of Perils Agricultural Issues May Be the Most Difficult to Solve." <u>Courier Journal</u> 2 Aug 1998: 20A.

Porterfield, Mannix. "W. Va. Ranked Among Nation's Top Polluters." <u>Register Herald</u> 11 Sep 1998.

Springston, Rex. "Efforts Start to Clean Up VA Rivers' Pollution Flows for 2,200 Miles." <u>Richmond Times</u> 11 Sep 1998, city ed.: B-1.

TIMETABLE

I intend to have all research completed by April 25. From April 26–May 3, I plan to work on my rough draft. May 4 and 5 I will use to revise and make corrections on my rough draft. I will turn in my final draft on May 6.

REQUEST FOR APPROVAL

I request your permission to write my report on fresh water pollution. If you have any suggestions or advice, I would appreciate your input. I look forward to discussing my topic with you.

Figure 2.2   A Typical Proposal Pattern

This template might propose a short research project in a college course.

PROPOSAL
(Title of Project)

To:
From:
Date:
Re: *(Optional subject line)*

INTRODUCTORY SUMMARY
(In this section you state your purpose—and your thesis, if necessary)

AREAS TO BE STUDIED
(Here, you discuss key points of your research. You can organize these in the order they will be covered in your forthcoming paper—that is, this section can be an out-line of your paper, using bulleted lists)

METHODS OF RESEARCH
(Here, discuss what research tools and materials are working for you. Follow with a source list in MLA, APA, or other standard style)

TIMETABLE
(Tell when you will complete the project)

REQUEST FOR APPROVAL
(Request permission to write the report and ask for any guidance the audience could provide)

In addition, content is introduced in two ways—as paragraphs supporting headings and as responses to questions asked in the proposal. Using subtitles to present the topic ideas of paragraphs is a standard approach in technical writing that eliminates the redundancy occurring if a separate topic sentence were to follow a heading and express the same statement as the heading. If these miniature titles carry the burden of declaring topic ideas one may skim the paper quickly for an initial understanding of its organization. This is exactly what a busy reader might do.

A similar organizational technique, the strategy of questioning, can evolve naturally from the preliminary outline of your proposal, since at the outline stage you may be querying yourself in a similar way as you try to define your project. In their general forms such questions can

◆ Define (What is X?)
◆ Establish background (How did X arise?)
◆ Show cause (What causes X?)
◆ Demonstrate effect (What are the effects of X?)
◆ Propose change (How can we solve X?)

Notice that these questions, in this sequence, can structure a short proposal. Also, if you are "stuck" while drafting your document, you can use these questions as prompts. Often, the final document's headings themselves are forms of these questions.

Note also that Amanda's proposal includes a list of resources, but that this reference list does not have to follow the other material in the document unless demanded by a format required in a particular discipline or specified by an instructor. Here, references are incorporated into the section that refers to them, "Methods of Research." Although some prefer the reference list to appear last, as it would in a longer document, Amanda has elected to end the prospectus with a request for approval, a good strategy in proposal writing.

The pattern employed is itself strategic:

◆ An Introductory Summary contains a statement of purpose explaining the intent and scope of the project.
◆ A section titled Areas to Be Studied divides up the project into the different concepts governing it, providing a "road map" of the project for the reader and writer alike.
◆ The Methods of Research component lets the reader know what materials the writer has consulted or intends to consult.
◆ The Timetable explains when the phases of the project will be completed.
◆ A Request for Approval provides a closing that encourages feedback from the reader.

Refer back to Figure 2.2 for a summary of this pattern.

Example 2. Proposals for Conference Presentations

What They Look Like.   A prospectus for presenting findings or reading a paper at an academic conference typically responds to a call for papers posted in professional journals or circulated to college departments. In general, such requests for proposals specify the appearance and content of the document that you will write. Since the conference coordinators must sift through mounds of these proposals, brevity usually encourages acceptance. Your prospectus could be written in letter form, but such format introduces extra elements into the proposal such as a closing and a salutation, and the logo of letterhead might itself consume needed space. Unless requested to do otherwise, you might achieve the best results with a one-page proposal in modified memo form. Any extra detail should follow as an attachment. This format works whether your proposal is produced as a printed page or sent electronically as an e-mail, since e-mail format is usually a modified memo form. Figures 2.3 and 2.4 are examples of such proposals.

Figure 2.3   A Proposal for a Conference Presentation

Proposal

For the 2003 Annual Meeting of the
Society for Scientific Aspects in Literature

TO: Phoebe Bernard
             School of Literature
             Peachtree College
             Atlanta, GA 30032

FROM: John Ingland
             Department of English
             Casey College
             P.O. Box 45
             Branton, VA 25899

DATE: March 10, 2003

Paper Title: "On the Verge of Empiricism—*The Tempest,* the Closed System, and the Assaying of Character"

This presentation shows how different views of the world coexist within this transitional play. Older images of hierarchy, correspondence, and organic interdependence share the stage with, but are not supplanted by, what we might call a more-recent "prescientific" concept. This second world-view, still a visual one, studies society and character within the metaphor of mathematical equivalence and the image of assaying, or balancing. *The Tempest*—itself a play about magic deriving from an overtly medieval lineage—is also a play about exploration, veering toward the discovery of "brave new worlds" and venturing towards a new empiricism. Its very rhetoric stands midway between multivalent poetics and an attempt to use language with figurative precision.

This presentation will take twelve to fifteen minutes and will require an overhead projector.

Figure 2.4   Proposal for a Conference Presentation

Proposal

For "<u>The Faerie Queen </u>in the Present" Symposium

TO: Elizabeth Tudor
              Department of English
              Whitehall University
              P.O. Box 8302
              New Hope, CT 06555

FROM: Burton Williamson
              Department of English
              Wasson University
              P.O. Box 679
              Bartley, CT 06444

DATE: August 18, 2002

The concepts of Plato, Pico della Mirandola, Sidney, and even Sebastian Brant shape the Renaissance discussion about art which provides a background for *The Faerie Queene*—a background fraught with ambiguities. My presentation will examine characterization and situation in *The Faerie Queene* against this context of the tension between good and bad art, artifice and nature.

I will first explore some conventional definitions of true and false art exemplified in the doings of Archimago and Acrasia, the Redcrosse Knight and Guyon. Life in the Bower of Bliss and the creation of the false Florimell will serve as expositions of Spenser's apparent valuing of invention over mimesis, an emphasizing of improvement upon nature over the world as it is or as it can be merely reassembled: the structuring devices of *The Faerie Queene* themselves attempt to promulgate "good poetry."

Yet, in attempting to adhere to a Sidneyan definition of good art, climaxing pictorially in the Mount Acidale episode, a self-reflexive uncertainty emerges for Spenser. For are not pastorals—the realm of Colin Clout—and romances—the stuff of *The Faerie Queene* itself—mere assemblages of conventions? Spenser must wriggle out of this paradox. In doing so, he both invokes neoplatonism and foreshadows a later debate over the nature of art.

This sixteen-page paper can be read in twenty minutes.

How You Do It. These somewhat fictionalized examples demonstrate the essentials required. You must indicate that you are proposing a paper or session for the conference. You must address the coordinator or reader who will be evaluating your prospectus. You must provide contact information about yourself. You must explain your presentation in a short abstract or summary and stipulate the resources necessary to present your talk. A common pattern looks like this:

◆ Heading material that identifies the conference and that provides contact information
◆ Paper or presentation title, if known
◆ Abstract of your paper or presentation (these are often collected and circulated among conferees, and may be single- or double-spaced as requested in the call for proposals)
◆ Indication of the length of the oral presentation
◆ Request for any special resources necessary (such as a computer projector or an overhead projector)

Again, be sure of the expectations of your particular discourse community and follow its rules about presentation—the suggestions above are general guidelines.

Example 3. Proposal Templates Integrated into Other Papers Having Problem/Solution Structures

What They Look Like. Multipart reports and research papers for class projects often contain elements of the proposal if these papers establish a problem, review ways to overcome the problem, and then urge action. Such papers may not be called proposals, nor may they be as detailed in establishing plans of action, but they clearly incorporate facets of proposal writing if they include the functions stated above. Here, for example, Figure 2.5 displays the table of contents in a research report by Samuel Deem.

Figure 2.5   Samuel's Table of Contents

Why He Did It That Way.   Note that this outline follows the pattern of a proposal. The paper begins with preliminary material—an executive summary and an introduction address the reader and explain the gist of the topic to be discussed. Next, Samuel divides up the paper into sections that correspond to the components of a proposal:

◆ *Definition and classification.*   First, lawsuit abuse is defined. This section also classifies the types of abuse, providing a taxonomy that the paper can use later.
◆ *Illustration.*   Next, examples illustrate the problem of lawsuit abuse.
◆ *Solution.*   Then, the paper discusses means of controlling the problem, including public-education campaigns and law reform.
◆ *Recommendation.*   A call to action summarizes the solutions and requests public response.

◆ *Conclusion.* The conclusion amplifies some of the data presented in the introduction, specifying the cost the public incurs as a result of frivolous lawsuits.

◆ *References.* Since this paper is written in MLA format, a Works Cited page ends the essay.

The design of the body of this report was the greatest challenge for Samuel, and the pieces of the paper did not coalesce into a unity until he found the right format for the job—which turned out to be the proposal format. "Starting the body of the document usually stuns me for a while," says Sam, but once understood, "the format is easy." Perhaps the proposal template will assist you, too, as you structure longer writing assignments.

What It Looks Like. Emily Montgomery was also seeking the best format for her research assignment, a problem-solution paper that explored the causes, the effects, and the prevention of juvenile crime. Emily was able to incorporate a cause-effect emphasis into a proposal pattern to create a clear structure for her project, as shown in Figure 2.6.

Figure 2.6 Emily's Table of Contents

TABLE OF CONTENTS

Executive summary	1
Introduction	2
Juvenile crime defined and its statistics	2
Causes of juvenile crime: what drives kids to kill	2
The deterioration of families	
Metabolic abnormalities and disorders	
Manifestations of juvenile crime	3
Behaviors exhibited	
Parenting interactions	
Prevention	4
What you can do	
What the government is doing	
Conclusion	4
Works cited	6

Why She Did It That Way.   Underneath its cover page, Emily's formal report begins with an executive summary—an abstract containing the gist of the presentation. That abstract also maps out the plan of the report so that even a quick reading of the abstract shows the report's structure clearly (see Figure 2.7).

Then an introduction follows (see Figure 2.8). Since Emily has just mapped the paper for the reader, she does not wish to repeat herself. Rather, she uses the introduction to alert the reader to the problem, and saves solutions for the recommendations section later in the paper.

Figure 2.7   Emily's Executive Summary

EXECUTIVE SUMMARY

This report is designed to inform the reader about the rise of juvenile crime. It also will demonstrate ways that parents and teachers can help to rebuild the lives of youths who exhibit delinquent behavior and will offer strategies to prevent this behavior from entering communities. My project surveys major areas of juvenile crime and consists of four sections:

1.   Juvenile crime defined

2.   Causes of juvenile crime

3.   Manifestations of juvenile crime

4.   Prevention

A source list at the end of this report provides access to further information.

Figure 2.8   Emily's Introduction

Introduction

Recent events have again focused the nation's attention on the violence in the United States caused by juveniles, an issue that has generated public concern and inspired research for more than three decades. The mix of children and guns has certainly reached a crisis in this country when gun use factors as a prominent cause of death among some teenage groups. Despite long-standing attention to the problem, it is becoming ever more apparent that increasing the number of police officers, intensifying prosecution, and expanding detention are not sufficient. Nor can they alone adequately deter juvenile violence and crime.

Emily next defines juvenile crime, but does not stop there. Rather, she employs this section to explain that although the problem can be defined, causes are not well-understood. This approach moves her discussion to the next stages of the paper, an explanation of causes and a review of the effects of juvenile crime. The recommendations section, labeled "Prevention," proposes a campaign of public awareness organized at the local and state levels. The conclusion is a final call to action, followed by a works cited page.

Emily was familiar with the principles of proposal writing when she wrote her paper. She notes that "I use proposal writing at work"—in the office of a physician—and says that "Knowing the styles and techniques of proposal writing has helped me substantially because it has given my documents structure and coordination. It allows my papers to flow and be easily read by others." She adds that "My greatest challenge is condensing my information. I always have a lot to say and by using the formats I've learned, I know exactly where my information needs to go." She recognizes the utility of the proposal template with its division of a paper into "departments." If you are uncertain about what you are to do in a given portion of the paper, and you are following a proposal model, you need only ask yourself what department you are in and what you are supposed to do while there. A research paper, then, may promote a position as does an actual proposal, and thus may be organized as a proposal. Similar rhetorical techniques apply in both instances.

Example 4. Proposal Techniques in a Prospectus for an Extended Project

It May Look Like This.   In applying to a program in cultural studies, Rachel Lanier found that she had to rely upon the elements of proposal writing in order to clarify her summary of intentions. She needed to explain her field of interest, discuss her intended project, and indicate her qualifications for such work. Reflecting on past experiences with writing proposals led her to design an essay using a modified prospectus model. The pattern she devised looked like this:

◆ Introduction of project
◆ Focus of project
◆ Design of project
◆ Qualifications of applicant
◆ Resume

A portion of the document produced by this pattern appears in Figure 2.9.

Figure 2.9 Portion of Prospectus by Rachel Lanier

Proposal: Project in Cherokee Pottery

Timeless, delicate, and beautiful. These are just a few of the words that could be used to describe the pottery created by the Cherokee people. For thousands of years, this form of art has been practiced by the Cherokee. And like the Cherokee, their pottery has undergone many changes throughout time. Despite these alterations, however, Cherokee pottery has remained a much treasured and practiced traditional art. It continues to tell the history and stories of the Cherokee people, forever preserving a link to the past and connecting past with present and future. My work in the cultural studies program will trace the changes in Cherokee pottery throughout history and examine why it changed.

Project Focus

Cherokee pottery, like other forms of Cherokee art, has evolved in many ways. The history of the pottery is a complex and varied one. As the focus of a degree project, this history will be broken down into four sections. The first will examine Cherokee pottery before colonization manifested serious effects on the Cherokee way of life. Pottery made during this time period was undoubtedly different because of the abundance of needed materials, less reliance on the settlers for goods, and less pressure on its makers to become "civilized."

The second and third sections discuss transition. The second phase will focus on how colonization affected the creation of Cherokee pottery. Because of the onslaught of settlers and the pressure to conform to the "civilized" lifestyle, the original practice of creating Cherokee pottery was lost. The third section will analyze how the Cherokee redeveloped their art of creating pottery through assistance from another tribe, the Catawbas. With their help, and in spite of the eradication of their art form, the Cherokee began to craft pottery once again.

The fourth and final section will focus on Cherokee pottery today. For many reasons, contemporary Cherokee pottery is different from that made thousands of years ago. This segment will explore how availability of materials, tourism, and mass marketing have all affected the way in which the pottery is fashioned and used. It will also examine the implications of these effects on the future of this traditional form of Cherokee art evolving under constant pressure from the dominant culture.

Project Design

The design of this project relies on directed independent studies and readings. Through them, I will collect the knowledge and research needed to underpin a thesis. These directed studies will focus on the Cherokee pottery created during each of the four time periods. Research on how the pottery is made, different designs employed, figures on the pottery, and the intended function of each kind of piece will supplement these studies. Another directed reading will focus on the general history of the Cherokee. These committee-directed projects will provide a good understanding of how the pottery changed and why.

Figure 2.9   *(continued)*

> To enhance the design of this program, additional elements should be added. Working with Cherokee pottery presents the opportunity to apply for grants for on-site research at the Cherokee reservation in eastern North Carolina. Grants from institutions such as the National Endowment for the Humanities could be used to cover the minimal costs associated with investigation of this type. The main goal of this on-site research would be to support the fourth section's analysis of Cherokee pottery today. These grants would allow extensive research to be conducted concerning how changes such as tourism and mass marketing have affected all aspects of the art form.

(Rachel then explains her qualifications in detail, covering her general coursework and her extensive grounding in Native American culture and literature. She closes with her curriculum vita.)

This example replaces the conclusion and reference sections usually found in proposals with depictions of the student's qualifications. Since the prospectus terminates in a resume, this resume can in fact be thought of as a "reference" page. Other aspects of the proposal template are familiar, though: introduction, definition or focus, and plan to be proposed. The introduction presents the subject to the reader, emphasizing its important qualities and closing by stating the mission of the project. The "focus" portion explains the project's four phases of inquiry. The "design" section discusses Rachel's plan: specific courses, assignments, and presentations that Rachel intends to incorporate, as well as grant funding she intends to seek. She has used the techniques of proposal writing to organize her thoughts and present them clearly to others.

Advice

Mission.   Every proposal needs a mission, without which it is ineffective. Often, as you write successive drafts of the proposal, you will clarify your mission. But be sure that the agenda of your project is absolutely apparent to the reader. In refining this mission, graduate and undergraduate students alike, seized by anxiety, sometimes hesitate to discuss their intended work with appropriate audiences before submitting a formal document explaining their projects. Don't fall into that trap. Your peers can tell you quickly whether your document communicates. Your professors are a valuable resource. Write in the language of your audience, and with the knowledge base of your audience in mind. Of course, it is likely that there will be an additional readership for your proposal. This might include fellow students, or a faculty committee. Study this audience's characteristics, too. If that audience requires more background information than your professor, then incorporate that extra material into the document.

Content. Such proposals must demonstrate your knowledge of content. In many fields, this knowledge is displayed in an extended review of the subject literature during a dedicated section of the prospectus. The types of such reviews and the degrees of their depth vary with the formal requirements in each field. You can be sure that your professor is familiar with the literature in the specialization under study, but since you must demonstrate that you have internalized that knowledge, you must be thorough. Extended reviews of literature as used in longer proposals appear later in this book.

Synthesis and Evaluation. Beyond demonstrating your assembling of knowledge, a longer proposal must reveal the ability to synthesize and assess such information. A research project resembles a house made of bricks and timber: an architect designs a unique house, but the bricks and timber come from manufacturers and suppliers. By comparison, you design the research project, though much—sometimes all—of the research derives from outside sources. Your design is your synthesis, your way of putting existing materials together in a new and innovative way using your own concepts. In reviewing the background of your subject, then, you would not discuss each of your sources in succession (as you took notes on them) but instead would organize your paper conceptually, employing different sources as needed to enhance your discussion of a concept. Similarly, you must often incorporate an evaluative component into a proposal, assessing the strengths and weaknesses of a source or project, and incorporating a holistic overview to do so. Since this aspect is a feature of longer prospectuses, it will be developed in a later chapter.

Protocol. Proposals cast content, synthesis, and evaluation into a format recognized by the discourse community one addresses. Items acceptable in one format may not be in another—for example, the use of APA citations in a humanities proposal would be a rarity. Granting bodies often specify formats that seem odd or repetitious to the proposal writer, but many such organizations "part out" the sections of prospectuses to different readers, necessitating duplication of material. What appears in this chapter, then, is a general guide to protocol, and should not be construed as applying in all cases. What DOES apply is the approach of technical communication, characterized by the organization of the document into headed topics, the use of parallel structure, and a reader-oriented philosophy.

For Further Study

1. What other aspects of academic life can involve the use of proposals? You might consider student organization projects, grant applications, requests to perform research, convocations, or graduate poster sessions.

2. Would another format of proposal, perhaps a letter, be best for certain tasks? For example, in some cases might a letter be the preferred means of communicating to an entity outside the college? Why or why not? How would you construct such a letter?

3. How could you combine a letter with a long proposal in standard report form?

4. If you had to give an oral presentation of your proposal, how would it differ from a written document?

5. How might you enhance your oral presentation with multimedia?

Review of Patterns

A Proposal Organized by Questioning Techniques

Introduce subject

Define issue (What is X?)

Establish background (How did X arise?)

Show cause (What causes X?)

Demonstrate effect (What are the effects of X?)

Propose change (How can we solve X?)

Follow with reference list

A Proposal for a Research Project

An **Introductory Summary** contains a statement of purpose explaining the intent and scope of the project.

A section titled **Areas to Be Studied** divides up the project into the different concepts governing it, providing a "road map" of the project for the reader and writer alike.

The **Methods of Research** component lets the reader know what materials the writer has consulted or intends to consult.

The **Timetable** explains when the phases of the project will be completed.

A **Request for Approval** provides a closing that encourages feedback from the reader.

(**References** may follow here rather than be incorporated into Methods of Research if appropriate.)

A Proposal for a Conference Presentation

Heading material identifying the conference and providing contact information

Paper or presentation title, if known

Abstract of your paper or presentation (these are often collected and circulated among conferees, and may be single- or double-spaced as requested in the call for proposals)

Indication of the length of the oral presentation

Request for any special resources necessary (such as a computer projector or an overhead projector)

Proposal Concepts Integrated into a Research Report

Executive summary

Introduction

Definition and classification

Illustration of problem

Review of solutions

Recommendation

Conclusion

References

Proposal Concepts Integrated into a Cause-Effect Paper

Executive summary

Introduction

Definition

Explanation of causes

Review of effects

Recommendations

Conclusion

Works Cited

Proposal Pattern in an Application to a Program

Introduction of project

Focus of project

Design of project (including timetable if needed)

Qualifications of student proposing project

Resume

CHAPTER

3

Further Lessons from Proposals in the Workplace

This chapter will examine some more workplace proposals, relating them to those inside the academy. Many lessons can be learned by studying the strategies of such documents and applying them to other, more complex uses.

Questions

◆ What are some more general paradigms for creating proposals in a business format?

◆ How can these be transferred to other proposals?

◆ How do I display content in these proposals?

◆ What patterns of organization are effective?

◆ What determines selection of format?

◆ Do these proposals have features in common with academic writing?

◆ What is an RFP?

Examples of proposals abound in non-academic settings. As we have noted, the commercial world is the site for many such items. In business, some are informal presentations and requests such as an employee makes to a supervisor in a meeting, pitching an idea and providing justification for it. A memo, letter, or e-mail might suggest a course of action to a co-worker or client, using a pattern and approach that we have studied in Chapter One and also when looking at short academic proposals. On the larger scale a report several pages long, such as the ISO review that we examined in Chapter One, might respond to a supervisor's question by recommending a course of action. Perhaps extensive documentation and support would dictate the use of a multisectioned proposal to get results. Such proposals are the subject of Chapters Four and Five.

To achieve results in the larger community outside the college, then, effective proposal construction certainly employs the writing strategies and formats that we have observed before, but with a difference. In the academic world one's audience is predefined, as the writer of a prospectus usually knows the needs and characteristics of the readership—a restricted group such as a class, committee, or board. However, the creator of proposals targeting commercial needs may have to research specific constituencies carefully, since a readership may not be clearly demarcated. Defining yourself and your prospective clientele becomes a major chore. In fact, you might have to create your prospective clientele through aggressive marketing.

Applications

Such proposal-writing can pose challenges similar to those of academic proposals, in that

◆ You must study your readership and craft the proposal to fit its sense of warrant, its underlying value system.
◆ You must accentuate the argument's outline so that its parts, used as headers in the text, are clear and flow logically.
◆ You must underpin discussion within the document with claims of fact, preference, and policy, using copious detail.

Remember too that generally, proposals of this kind divide into two classifications, those seeking agreement and those calling for action. For example, a report demonstrating that something is feasible, that it can be accomplished, seeks understanding and assent. But a document may also move beyond proving that something is feasible, by stating that it is the best policy—and may then urge action to implement that policy. We will look at several examples of documents that depend on the proposal patterns of establishing feasibility or calling for action.

Note that such documents may be self-initiated. Or, they may be constructed in response to a notice or advertisement called an "RFP," a request for proposal. If the prospectus is self-initiated, it should result from careful research of the audience or facility that is the intended subject of the document. Its writer will have to decide on the placement and labeling of key elements, borrowing format from comparable proposals. Of course, a proposal responding to an RFP has the advantage of stating its case within the format prescribed by and to the audience specified in the RFP. Such a request for proposal may define the readership, scope, subtitles, topic sequence, and style of the expected document. An RFP certainly will require the respondent to

◆ provide information within its stated parameters
◆ adhere to a posted deadline
◆ mail or transmit the proposal to a contact person

Additionally, RFPs often include the name, telephone number, and e-mail address of someone who can assist the respondent by answering questions about the project. Documents that vary from the design specified by the RFP are unlikely to receive appropriate attention. Though RFPs announce grant and bidding solicitations, it is unusual for an RFP to be posted inside a company or agency. And the RFP differs fundamentally from a simple call for bids. Bids are taken for approved projects; a proposal seeks approval for a project and may, in fact, suggest a project that had not been contemplated before.

Example 1. An Internal Transmittal Serving as a Short Proposal

What It Looks Like. Figure 3.1 presents a document written not as a letter, memo, or report but instead as a transmittal—a form to be filled out by an employee who will submit it to a supervisor in charge of budgetary requests. The workplace uses such transmittals to achieve clarity and efficiency, since they are premade templates designed to make presentations and the reviewing of presentations uniform. There are many versions of these documents. Often, they are files available online or on disk, designed to be filled out by personnel. Of course, the location of the information in such a template varies with the internal styles of each company. But such communications are employed throughout business, agency, and academic settings. The forms may be used to request *capital* expenditures for a major piece of equipment, for renovation of a facility, or for some other large form of expansion or development within a corporation, agency, or institution. Alternatively, similar forms request routine *operating* expenses necessary to sustain ongoing activities.

The document in Figure 3.1 is written by Neil Manning, a project development engineer since 1992 and a former army engineering officer. I interviewed Neil about how proposal writing functions in his job.

Q: What is your line of work?

A: Product development for a magnetic materials manufacturing company.

Q: How do you use proposal writing at work?

A: I use proposal writing for capital appropriation requests and occasionally to pitch new projects.

Q: Has knowledge of this style and technique helped you?

A: Yes, it helps organize my thoughts and more objectively evaluate the request.

Q: What is your greatest challenge when constructing a proposal or similar document?

A: Writing about engineering topics to non-engineers. Particularly, shifting the emphasis to financials.

Q: What tips can you offer someone learning to write in this style?

A: Try to think like an accountant—it's easier for an engineer to do this than for an accountant to think like an engineer.

Q: What is the basis for the particular item you have contributed?

A: A critical piece of support equipment was getting too old to fix economically. We had the choice of replacing it or using outside services.

Q: What part of the document was easiest to write? Why?

A: The financial data was easy because most of it came from formal quotes.

Q: What was difficult, and why?

A: Identifying enough financial reasons why the money should be spent and stating this so that purchasing the equipment is the logical choice.

Q: Are there any features of the document that you would like to emphasize for us?

A: The document leads you through the process. Since our documents are templated in software, dumb math errors are eliminated. Hopefully we will get these proposals approved electronically in the near future so the paper can be eliminated too.

Figure 3.1 Capital Allocations Proposal, Courtesy Neil R. Manning

My Co. Incorporated			DATE 11/23/99		PROJECT NO.			
CAPITAL APPROPRIATIONS REQUEST			DIVISION "The R&D Engineering Co."		LOCATION Somewhere, USA			
PROJECT TITLE Carbon-Sulfur Analyzer			BUSINESS SEGMENT R&D					

PRIOR	TOTAL: PRIOR APPROVAL AND THIS REQUEST	DESCRIPTION	AMOUNT OF THIS	FORECAST OF CAPITAL EXPENDITURES BY ESTIMATED TIME OF EXPENDITURES I:							
APPROVAL			REQUEST	4th Qtr 99	Qtr	Yr	Qtr	Yr	Qtr	Yr	Qtr	Yr
		LAND AND BUILDINGS									
	54,775	EQUIPMENT	54,775	54,775							
	0	RELATED EXPENSE	0								
$0	$54,775	TOTALS	$54,775	$54,775	$0		$0		$0		$0

DATE OPERATIONAL	DATE FULL PRODUCTION		EFFECT ON ANNUAL PROFIT (IN THOUSANDS OF DOLLARS)				PAYBACK PERIOD AFTER TAXES
YEAR OF OPERATION		1st YEAR	2nd YEAR	3rd YEAR	4th YEAR	5th YEAR	6-10 AVG.	YEARS
PROFIT AFTER TAXES								DISCOUNTED CASH FLOW
RETURN ON TOTAL INVESTMENT (AFTER TAXES)								%

PROJECT CLASSIFICATION	[ ] COST REDUCTION	[ ] EXPANSION	[ ] SAFETY & ENVIRONMENT
BUDGETED: YES ( ) NO (X)	[ ] PROFIT MAINTENANCE	[ ] NEW PRODUCT	[X] OTHER

SUMMARY OF PROPOSAL:

The present Ledo Carbon-Sulfur Analyzer is wearing out. The control console's printer (which uses a cash register type of paper) is not printing clearly. The analyzer's furnace has been tripping its breaker repeatedly in the last year. We have had servicemen attempt to repair the unit, but the intermittent nature of the fault has made it all but impossible to diagnose, let alone correct. The unit was purchased sometime after 1984 and has been used primarily to determine the carbon and sulfur levels of the special alloy ingots prior to processing them. Without this instrument, melt samples would have to be analyzed outside the plant at significant expense and inconvenience.

Very Expensive Laboratory, Inc. charges: $50/sample at 700 samples/year = $35,000.
   Turn around time: 24 hours.
Kind of Marginal Analysis Co. charges: $14.99/sample w/next day FAX (72 hrs. of rec't. of samples) (Reg:$24/sample);
   (~$10,493/year).
Additional costs: Cost of Purchase Orders and handling to ship the samples.

It is desirable to analyze the materials in house because of timing and storage requirements for the ingots. Additionally, the equipment is used for analyzing other products, such as incoming material quality control.

PLANT APPROVAL		DIVISION APPROVAL		CORPORATE APPROVAL	
INITIATOR	DATE	V.P. TECH.	DATE	SERVICES	DATE
MANUFACTURING ENGINEER	DATE	V.P. MATERIALS	DATE	GROUP EXECUTIVE	DATE
MANAGER CUSTOMER SERVICES	DATE	V.P. SALES OR V.P. HUMAN RESOURCES	DATE	FINANCE	DATE
GENERAL MANAGER	DATE	V.P. CONTROLLER	DATE	PRESIDENT	DATE
MANAGER FINANCIAL ANALYSIS	DATE	DIVISION PRESIDENT	DATE	C.E.O.	DATE

(continued on next page)

Figure 3.1 (continued)

ITEM NO.	DESCRIPTION	NO. OF UNITS	CAPITAL COST	BASIC EQUIP ITEM	INSTALLA-TION	FREIGHT	RELATED EXPENSE
1	Ledo CS-40 Carbon-Sulfur Determinator	1	$ 47,310	$ 47,310			
2	Operating Supplies Kit for Ledo CS-400 Carbon-Sulfur Determinator	1	$ 759	$ 759			
3	Autocleaner/Dust Removal Kit	1	$ 2,995	$ 2,995			
4	Recommended Spare Parts	1	$ 1,103	$ 1,103			
	Contingency (5%)		$ 2,608				
	TOTALS		$ 54,775	$ 52,167	$ -	$ -	$ -

PREPARED BY

PROJECT TITLE

APPROVED BY

SHEET NO.   OF PROJECT NO.

CAPITAL PROJECT DETAIL
MACHINERY AND EQUIPMENT

40

How It Works. This document is designed to be filled out easily by the proposer and read at a glance by the person in charge of the budget. The proposal consists of two parts: a capital appropriations request for funding, and a worksheet detailing items that correspond to lines on a budget. Though these sheets are designed to improve the paper flow within a company by eliminating the need for an added memo summarizing the proposal, a covering memo might accompany these sheets if the request were not an ordinary one for Neil. The pattern then would be:

◆ An explanatory memo, if needed
◆ The appropriations request, containing a rationale for the proposed purchases in the space marked "Summary of Proposal"
◆ A budget worksheet showing the items to be purchased, their cost, and other related expenditures that can be expected

Inside the proposal there are a number of considerations worth reviewing:

◆ Initial detail and justification appears in the heavily-formatted boxes that constitute the "introduction." This part even contains a table in case one must demonstrate how the expenditure can be recaptured.
◆ The summary section begins with a concise explanation of the problem requiring attention and an explanation of why the problem needs to be solved in the suggested fashion.
◆ Justification continues in a new paragraph, and includes a cost analysis of alternatives to the proposed solution.
◆ A statement condensing the discussion follows. That statement elaborates on the reasons to follow the requested action.
◆ Both sheets include sign-off areas for supervisor approval. In the interest of efficiency, this internal communication forgoes the elements of closure that might be expected in a sales proposal, and the approval sections instead serve that function.

The transmittal's elements, then, correspond to those of more traditional proposals. For example, budget allocation forms will often accompany longer proposals that present intended expenditures. This is especially true of grant proposals written to the specifications of the grantor and containing a matrix in which to put your anticipated budget and expenses. However, the layout of such a tabular document, with its constraints on space, can promote a terse, "telegraphic" style of writing that, though it has a place in the realm of forms, should, of course, be avoided in more conventional proposals. Paying attention to what is acceptable "house style" in your workplace will assist you; communication in some disciplines such as engineering or the health fields tends to use short phrases that are universally understood within the discipline, whereas writing in the humanities frequently involves inventing and then explicating terminology.

Example 2. A Sales Letter

What It Looks Like.   The sales letter's purpose is to achieve action, re-sulting in a contract that will generate revenue. The parts within the letter constitute a basic outline of the presentation, as we have seen in examining other proposals. An understanding of those receiving the communication is crucial, as the letter should not repeat the obvious to knowledgeable readers if these are the audience, nor be beyond the understanding of general readers, if these constitute the audience. An example of a sales letter appears in Figure 3.2. It is generated as part of a stream of continuing external corre-spondence, and not as a response to a formal request for a proposal.

Figure 3.2   Sales Letter

Schmendrick Brothers Instruments
921 Sourwood Ave.
Creve Coeur, MO 65813
1-314-555-5555
schmendri@instrum.com

Ray Ramirez
41 Raleigh St.
Halton, MO 66152

Dear Mr. Ramirez:

Attached is your check in payment for the L48 guitar and case, serial num-ber/factory order number Z-8553-33. The instrument arrived in excellent shape and our client will be pleased with it. We are still interested in exchanges of merchandise as discussed before, or in selling a number of pieces to you at below-market prices. I propose that you purchase our sur-plus inventory at cost, or exchange it for like inventory at cost.

In our stock, for example, we find the following items that would benefit your operation rather than ours:

   1960s Framus three-pickup hollowbody electric, blonde finish, double-cut-away, jumbo frets;
   "O" size Lyon and Healy parlor guitar, grained maple body, red spruce top, pyramid bridge, bound top;
   "Troubador" guitar circa 1900, birch, stained brown, maple fingerboard;
   Stella-made guitar circa 1930, refinished checking on top;
   Early 1900s parlor guitar with fancy purfling, chip fixed in soundhole, body reglued. This guitar has been rebraced with x-strutting, and would take very light steel strings. The sides and back are maple and stained dark brown;

Figure 3.2   *(continued)*

P'Mico "Nobility" archtop guitar. Fancy top purfling, maple or birch
   throughout, dark sunburst, replaced tuners, old alligator-style case;
Concertone tenor banjo, maple neck with some flame, small steel tone
   ring, maple rim, heavy hardware, bound neck, birdseye resonator;
Gusli and bow, small and made for a child. The instrument has beautiful
   chip-carving on back.

These are merely samples of the approximately two hundred instruments
accumulated in forty years of buying from estates and owners. We will e-mail
you a complete list upon request, and are attaching sample pages of our
inventory list, with net pricing, for your review.

Please let us know immediately which items appeal to you. If we exchange,
we are, of course, interested in trading for other instruments with main-
stream brand recognition. Please contact us by telephone or e-mail so that
we can work out the details of an inventory exchange or purchase.

Cordially,

Alfred Schmendrick

Alfred Schmendrick

enclosure: sample pages of inventory list

How To Do It.   The writer, on a mission to increase business liquidity, uses a
full block style to impart a sense of directness in his dealings with Mr. Ramirez.
He could employ a simplified style that contains a memo-like subject line, but
that mode of correspondence might be too abrupt for the audience, and
might suggest that the communication is a quick modification of a standard-
ized form letter. Moreover, this correspondence is brief anyway, as the writer
has kept the document small enough so that the reader can immediately identify its purpose by surveying it. Sometimes in business writing, of course, you
will see the salutation used to identify the subject or purpose—as in "Dear
Franchisee"—but this approach can also backfire since it is not a personal
greeting. Moreover, one could title the proposal right below the letterhead, as
is done in some advertising. The writer of this letter would not go that far—
he wants to convey a personal tone. He is also careful to impart "good
news"—payment for a previous transaction—at the outset, and to remind the
reader of a previous conversation exploring what the letter proposes.
   The plan of any sales letter should be visually-pleasing. Little is gained by
stacking densely-written, thick paragraphs on top of each other. Short paragraphs and indented lists should predominate. In a longer format, these
might benefit from headings employing key phrases or questions that define,

establish background, show cause, demonstrate effect, or propose change. Visual relief is so important in convincing the reader that the unsolicited letter without that characteristic is likely to land in the garbage can, unread. Be sure your document promotes eye movement and varied appearance of text. Revise it until it has some or all of these aspects:

◆ Visual balance, achieved by centering the text on the page so that white space does not predominate below or above it
◆ Major and subordinate headings introducing the reader to important ideas and demonstrating their hierarchy
◆ Indented lists of subordinate material presented using parallelism
◆ An extra line of spacing between the single-spaced paragraphs in the text
◆ A type size and font that are easy to read
◆ Distracting subordinate material presented as an enclosure, not located in the main text
◆ Brevity in the text of the body

Note that the letter of Figure 3.2 includes a "teaser"—a selection of appealing examples inserted to entice the recipient into agreement. This letter also incorporates a request for action into its closing element. Here there is the implication that Mr. Ramirez must contact Schmendrick immediately lest choice items "get away."

The sales letter's pattern also follows standard proposal practice, in that

◆ An introductory paragraph explains the message.
◆ A section in the body discusses the different concepts associated with the topic of the letter.
◆ A following section elaborates.
◆ The letter presents a time frame for action.
◆ A request for approval in the closing asks the reader to respond.

Example 3. Proposal for a Contract

What It Looks Like. This is my sales prospectus for the book you are reading (Figure 3.3). The proposal itself was accompanied by a cover letter in sales format, contacting the acquisitions editor and addressed to him personally. It was not a "solicited" proposal, but rather one that developed from suggestions by my colleagues promoting such a project. Had it been a response to a request for proposals it would have followed the outline and content specified by such a request. The entire prospectus could have been written in sales-letter form, but I felt that doing so might add distracting elements to the presentation. I included as attachments to the proposal an updated vita or extended resume, and an analysis of other books overlapping my subject area. These were attachments because in business writing, any extra detail usually follows the main document rather than being incorporated into it.

Figure 3.3 Sales Proposal

Text Prospectus:
Form and Content: Proposal Writing Across the Disciplines

To: Steve Helba, Editor, Prentice Hall—Technology
From: Brian R. Holloway
Date: 8 October 1998
Re: Cross-Disciplinary "Proposal Writing" Text with Disk

Summary
I wish to write a short textbook presenting and analyzing proposals used in different disciplines—sales, engineering, science, healthcare, the humanities. A disk accompanying the book will contain the templates necessary to construct these proposals, thus providing the student and the employee with easily-understood models to emulate.

Rationale
College English classes rarely teach the art of proposal writing, necessary for the workplace and for graduate school. In addition, typical one-semester technical writing courses do not have time to discuss the proposal in depth. To respond to the needs of advanced undergraduates and graduate students, many schools now offer "professional writing" or "writing in the professions" courses. Yet texts for such classes, unfortunately, are few. To complicate the problem, most "multidisciplinary" texts for "advanced" writers feature a humanities orientation and present articles <u>about</u> the disciplines rather than articles <u>from</u> the disciplines—thus ensuring confusion for students planning to enter health, business, engineering, science, or the social services—and frustrating many undecided students, unsure of the requirements and writing protocols of different fields.

In most fields, however, a student must write a proposal—be it a clinical, business, grant, or graduate research prospectus. A book teaching the craft of such documents must

- **survey many kinds of proposals,**
- **present models of papers in the different disciplines,**
- **provide analysis of these models.**

(continued on next page)

Figure 3.3 *(continued)*

Approach/Pedagogy

A practical text will include examples and analysis, and be organized according to rhetorical type. Short documented proposals from the sciences, engineering, humanities, business, and medicine will serve as models for emulation and discussion. After each paper, a statement supplied by the paper's author will explore the key challenges and strategies of the document. My comments, focusing on style, structure, and format, will follow. Finally, exercises will ask students questions about these documents. This book, then, emphasizes **techniques**, **structures**, and **protocols** that work.

The text will function as

1) A book to **teach from**, to use as a reference for the protocols of many types of prospectus writing (even academics may not know the requirements operating outside of their fields);
2) A book to **complement** a handbook of style, usage, and documentation patterns without being such a text itself.

Educational Pragmatism

The models in this text represent documents similar to those that students in different fields will have to write. The student learns by studying and emulating real examples. Exemplary student work included in the text also reinforces the feeling that such tasks are achievable. The text's pragmatic approach derives from my twenty-three years in teaching, my twelve years in business, and from discussions with engineers, scientists, and executives.

Goals

The text has four main goals:

1) **Motivating** students who will see the relevance of writing in their fields instead of having to endure generic documents presented in typical texts;
2) **Exposing** students in a heterogeneous class to the multidisciplinary experience of reading proposals outside their chosen disciplines and commenting upon those papers;
3) **Teaching** the fundamentals of proposal writing using appropriate models;
4) **Providing** appropriate models for such writing in the workplace.

Figure 3.3 *(continued)*

<u>Coverage: Contents</u>

Structural and stylistic conventions

- Style
- Audience
- Tone
- Structure
- Format
- Documentation
- Questions

Short forms: semiformal documents from different fields

- Investigative documents
- Persuasive documents
- Questions

Formal documents in different disciplines

- Experimental and engineering proposals
- Feasibility assessments
- The humanities proposal
- The grant prospectus
- Larger technical and business proposals
- Questions

<u>Specifications</u>

General Description

I'd suggest that this text would work best as an inexpensive softcover book 6 inches by 9 inches, ring-bound or comb-bound so that it will open flat. The brightly-colored cover should be of durable card stock. A one-color design for the text could utilize gray screens of different densities. Length should be about 200 pages.

Manageable size, convenience of use, and affordability will help market the book.

Figure 3.3   *(continued)*

Visual Features

This is a visual book. Line drawings will be needed to demonstrate development and research strategies as well as to depict graphical material in the various papers and articles used. In addition, the papers (or parts thereof) must appear as "figures" to be rendered authentically. Depending on "house" design and manuscript preparation process, the number of such figures might be large.

Personal Qualifications

I helped create my first multipart professional document in 1973 and 1974, when, as an undergraduate, I joined a committee developing a text and revamping a course in the College of Education, University of Missouri. I earned both B.A. and M.A. in English from the University of Missouri, and a Ph.D. in English Literature from the University of Illinois, teaching sections of rhetoric for engineers, rhetoric for advanced students, and advanced expository writing—in addition to a variety of literature courses. At Parkland College, I taught business-oriented writing, along with a full complement of other composition and literature classes. I also served as a director of a retail corporation from 1981 to 1993. I am now Associate Professor of English at The College of West Virginia, teaching courses in interdisciplinary studies, composition, technical writing, and literature. This Spring, I will teach a new graduate-level professional writing course that I have developed. I have received both student government association and alumni association awards for excellence in teaching, have published a number of articles, and have written <u>Technical Writing Basics: A Guide to Style and Form</u> (Prentice Hall). I've attached a copy of my curriculum vitae.

Development Schedule

I estimate a minimum one-and-one-half-year development time to create the text, solicit and integrate the documents, and acquire permissions for necessary items. Please call me at 555-777-0000 (home) or 555-777-7777 (work) if you have any questions about my proposal—or e-mail me at **name@university.edu** if that is more convenient. I look forward to your reply.

How You Do It. This proposal for contract demonstrates the essentials required. You must address the reader in a heading. Early in your document you must find a warrant—a common ground upon which you and the reader stand. In this case, the understanding is that both writer and reader have noticed a problem requiring solution—a challenge. Once that understanding is acknowledged, the proposal can then discuss the solution using supporting details. Here is how the proposal works:

◆ *Audience address.*   A header in memo format acknowledges the audience and, in its subject line, states the focus of the proposal.
◆ *Summary.*   An opening summary provides a statement of purpose.

- ◆ *Rationale.* This section explains the reason for the project, and ties that to the warrant shared by the writer and the reader of the proposal.
- ◆ *Approach.* This section explains the function of the proposed product. The paragraph following it, *"Educational Pragmatism,"* reinforces the philosophy behind the product. These parts correspond to the *Definition and Classification* sections of some proposals.
- ◆ *Goals.* This section explains what will be accomplished by the product.
- ◆ *Contents.* This part explains what will constitute the product.
- ◆ *Specifications.* The paragraphs here show what the product will look like. This section contains discussions of cost. The *Contents* and *Specifications* sections correspond to the *Illustration* or *Examples* sections of some proposals and also provide detail about the solution that the product represents.
- ◆ *Personal Qualifications.* A section such as this also appears in many grant proposals.
- ◆ *Development Schedule.* A schedule of development reinforces the idea that the project is feasible.
- ◆ *References.* References and details of personal accomplishments are provided in the extended resume that follows this document.

This prospectus packs much material into a small space and conforms to the expectations of its readership. Variations on its pattern will occur in different fields because each particular discourse community will have its own rules about presentation.

Example 4. Research on a Workplace Topic: Integrating Material

What It Looks Like. Teresa Massie-Workman researched a topic of social concern as well as of vital interest to an employer: the effects of domestic violence on female employees. The introductory matter in her proposal covered a number of issues affecting such women, including health problems; welfare legislation stressing employment at marginal jobs instead of improving education to escape abusive situations; lack of child care facilities; workplaces ignoring the problems affecting these employees at home. A section of the introduction appears in Figure 3.4.

Figure 3.4 Part of Introduction: Sections of Proposal by Teresa Massie-Workman

> For many years now, researchers have documented the incidence of male abuse directed at women in their intimate relationships. The majority of this research has focused on the effects of violence on such women's psychological and physical safety. There is very little research about how abuse affects females' participation in the labor force. The purpose of this study is to identify perceived barriers to obtaining or maintaining employment experienced by victims of domestic violence.

Figure 3.5 Defining a Term: Sections of Proposal by
Teresa Massie-Workman

DEFINITION
    The definition of "domestic violence" varies throughout studies and is
often used interchangeably with that of the terms "intimate violence" and
"family violence" (Gelles, 1990). The majority of definitions discuss conduct
such as physical force, mental abuse, and control including intense criticism,
verbal harassment, sexual coercion or assaults, isolation due to restraint from
normal activity or freedom, and denial to access of resources. Increasingly,
the definition of domestic violence incorporates all intimate relationships, not
just married couples (Laurence and Spalter-Roth, 1996).
    Domestic violence follows women in every aspect of their lives, includ-
ing work. Measuring the depth of this problem has proven to be difficult for
several reasons. First, many women often hide or deny the abuse out of fear
that the exposure will have a negative effect on their employment. Secondly,
studies have shown that few employers know the signs of abuse, or take the
time to investigate the problem. This leads to a cycle in which the employers
see little need to respond to a problem that only occurs occasionally and the
victim sees little reason to reveal her abuse in an environment that chooses
to ignore the issue.

The proposal presented this material using a short introduction resembling an executive summary, followed by a definitional section (Figure 3.5) en-
abling readers to understand exactly what Teresa meant by "domestic violence." In the definitional portion of the document, Teresa had to confront and
explain the meaning of the key phrase "domestic violence" so that it would be
clear. She opted for several techniques of definition that could inform the
reader without deflecting attention from the proposal.

How Definition Works. In this context, elaborate definitions of terms are
not necessary unless the proposal presents technical information for a general
reader. Sometimes tables of definitions or abbreviations will appear as back
matter. Here, the writer needed clarity and included a definitional section in
the body of the proposal, listing behaviors generally agreed to be synonymous
and explaining the changing boundaries of the definition. She illustrated the
features of those affected by domestic violence. Then, she discussed the connection between such violence and its impact on the workforce. Since this
proposal focuses on the workplace, everything—no matter how theoretically-
based—must return swiftly to the practical issues. All segments of this proposal must reinforce recommendations addressing need: what can the employer do? Of course, one can see that this kind of writing can make an easy
transition from the specific issues of employment to the more general social
context motivating community betterment.

Advice

Warrant. The action-oriented proposal will not work if its readers and its writers have, or appear to have, different conceptual and value systems governing how they perceive the world. That is one reason for Teresa's definitional section—to reaffirm understanding. Stated another way, the reader and writer must embrace the same warrant for the document to succeed. Advertisers spend millions of dollars researching warrants so that their clients can communicate with their intended markets. Cultural differences are especially important and often misunderstood. For example, a practicing traditional member of a native American tribe tosses a cigarette from his car window to leave an offering for the spirits of those killed in a roadside accident. His warrant motivates an action that is poorly understood by the policeman from the dominant culture who pulls him over and begins writing a ticket for littering. As we embark upon a new century with its unparalleled social diversity, we need to be aware of a diversity of warrants.

Claims of Fact. Facts themselves do not constitute claims of fact. The claim of fact is a general statement that data must support. Normally the claim of fact appears in a heading or within the topic sentence of a paragraph. The paragraph or paragraphs that follow contain the material justifying the claim. Claims of fact often change as new data become known or existing data are reinterpreted. The monk of 1300 CE who asserted that the Earth was stationary and that the Sun revolved around it made a claim of fact, which he would have supported with a "common-sense" example—everyone could see the Sun rise in the morning and set in the evening, so it must orbit the Earth. But the great historical changes in the history of science are accompanied by reinterpretations giving rise to new claims of fact.

Claims of Preference. The claims of preference, or value, assert that we should prefer one thing over another. These, too, must be bolstered by appropriate data and justification. If pizza number two really does have more pepperoni and thicker tomato sauce, and if that is what we value in a pizza, we might find its advertiser's claim of preference justified.

Claims of Policy. The claim of policy drives the call to action or the closing recommendation of a proposal. Its statement is usually governed by this formula: because we value certain things, and because there is a challenge to overcome in maintaining our values, we need to implement a particular policy. For example, because we value our children's well-being, and because they need a place to go to keep them out of trouble, we should seek public funding to turn the abandoned market into a community teen center. Because you value economical gasoline use, and because your old car cannot fulfill this requirement, you should trade it in on a new, shiny, economical . . . well, you get the idea.

For Further Study

1. What claims of fact, preference, and policy appear in your own proposal writing? How have you answered the needs and knowledge level of your readership?

2. What can you do in your own writing to ensure a commonality of warrant so that you and the reader of your proposal proceed from the same system of values?

3. How will you handle the challenge of inserting needed research information into a "practical" proposal?

4. What similarities exist between sales-oriented and other types of proposals?

Review of Patterns

An Appropriations Request

Explanatory memo

Appropriations request and summary

Budget worksheet

A Sales Letter

Introductory paragraph explaining the message

Body discussing the different concepts associated with the topic of the letter and elaborated in subsequent sections

Time frame for action

Request for response

A Sales Proposal

An executive summary provides a statement of purpose.

A rationale section explains the reason for the project.

An approach section explains the function of the proposed product.

A goals section explains what will be accomplished by the product.

A description section explains what will constitute the product.

A specifications section shows what the product will look like and cost.

A schedule of development reinforces the idea that the project is feasible.

References, details of past accomplishments, and marketing reports are among the items that may be included as back matter.

CHAPTER

4

Crafting the Long Proposal— Some Examples

This chapter will examine organization and style in formal proposals. It will show how theoretical and applied features interconnect in this kind of technical writing that unites the concerns of workplace and academy.

Questions

◆ What are the organizational features of formal proposals?
◆ What stylistic considerations must the writer of such documents know?
◆ How do I present content within the sections of these documents?
◆ What do these proposals borrow from the technical report?

Examples of formal proposals include the article-length document and the thesis prospectus. You might initiate such documents themselves by proposing them in informal presentations that use the pattern of the short academic prospectus. These larger writings rest on extensive support and argumentation developed within a structure determined by the requirements of your discourse community.

Of course, because of the considerable variation in these requirements, you should not assume that one version of the large prospectus applies to all situations. For example, if you are in graduate school, your particular field, program, and committee requirements might dictate the construction of a document quite different from any in this book. What this text can provide, however, is a sense of the overall philosophy of creating the prospectus and an understanding of the requirements of style, audience address, content, and rhetorical underpinning of such writings.

Applications

Both formal proposals and their shorter relatives require that you

- Adjust your prospectus to reflect the vocabulary, stylistic expectations, and value system of your readers.
- Use generous spacing, transitional headers, and other features of technical writing to promote clarity.
- Follow a modified version of the template for technical reporting.
- Perform meticulous research to bolster and enhance all discussion within the document.
- Plan ahead so that the proposal itself can become the blueprint for the article or longer work that you intend to produce—sort of a super outline.

In such proposals, the text must inform and persuade. For the persuasive pattern to operate, the informational function of the proposal must operate.

Patterning in the Large Proposal

The parts that are briefly-stated in short proposals are likely to expand in size and function in a major prospectus. You will see these parts and their roles in the major documents presented in this section, which contain annotations describing them. In addition, these parts will relocate themselves depending upon the requirements of the prospectus. When examining such proposals, generally you may find variations on such templates as those listed below.

Experimental Project. The classic proposal for an extended activity based on experiment may follow a pattern such as this one, which has a number of departments that answer questions of equipment, physical space, theory informing the tasks, and other issues requiring commentary. The sections themselves may be headed differently than described, but may perform the same functions as those delineated below:

- Introduction—surveys the area of study broadly and then transitions to the specific purpose of the project. Many introductions work by a process of "genus-differentiae," first showing the bounds of the area and then distinguishing the way in which the proposed research topic differs from what has been examined before.
- Statement of purpose or problem—depicts the specific issue or question and discusses how the research question will be addressed.
- Rationale—summarizes the theoretical underpinning of the project itself. This section is important to grant and organizational proposals, too, and may appear in a different location in those documents—frequently earlier in the text.

◆ Hypothesis—explains what is expected to occur.
◆ Stipulations—these limit the role of the proposed project, providing an explicit boundary beyond which the project will neither operate nor from which the project will seek to generalize.
◆ Definitional section—this explains key phrases and abbreviations. This section may appear in a different location depending on the protocol of the prospectus and field. Some proposals for grant funding require this section to appear earlier in the document.
◆ Literature review—surveys the research on the topic.
◆ Discussion of method—shows how the project will be done, and what equipment, design, ingredients, evaluation and analysis will be performed.
◆ Timetable—may be incorporated into the "method" section instead, or may appear in an appendix—often in chart form.
◆ Appendix including certifications of compliance—a given in the case of any human-subject research or work with any hazardous materials.
◆ References—in the style appropriate to the field. Some proposals for grant funding require a different location for, and labeling of this section.

This pattern that structures experimental proposals can be used as a basis for constructing a prospectus for qualitative, rather than quantitative projects.

Project Relying on Qualitative Research. Such projects may depend on interviews, surveys, and Internet responses to generate primary information. The prospectus for a qualitative program may follow a pattern similar to the prospectus for an experimental one:

◆ Introduction
◆ Significance of project—provides the rationale of the project.
◆ Role and scope of project—functions to stipulate what will and will not be done.
◆ Definitions
◆ Literature review
◆ Methodology—explains the what, the who, the how of the project. What will be the design of the research project? How will information be gathered?
◆ Expectations—how will the data be presented and to what use will they be put? What findings will occur, and what will be the implications of the study?
◆ Timetable—might be incorporated into the methodology section.
◆ Appendix—including certifications of compliance.
◆ References

As you have no doubt noticed, both qualitative and quantitative proposals are not too different from the general models employed outside the academy.

Simplified Model.   For example, many prospectuses follow the general outline below, a plan resembling institutional or business templates, as shown:

Prospectus for an Academic Project	**Business or Institutional Proposal**
Abstract (if required)	Executive Summary
Introduction	Introduction
The Problem	Statement of Challenge
Hypothesis	Statement of Agenda
Project Method	Means to Address Challenge
Expectations	Expected Results
Timetable	Timetable
Appendix	Supporting Material
References	References

Or, as in the following examples:

Abstract or Summary	Executive Summary
Introduction	Introduction
Rationale	Justification of Project
Methodology	Means to Implement Project
Staffing	Personnel Required
Budget	Budget Needed
References	Supporting Materials
Appendix	

When surveying these patterns and possibilities, remember that when the
research proposal is integrated into a technical report template (see Appendix) it will acquire the usual apparatus of the technical report, including
an abstract or executive summary, table of contents, and so forth, resulting
in a format such as this:

◆ Cover with label
◆ Title page
◆ Abstract or executive summary
◆ Table of contents (and list of illustrations, if necessary)
◆ Introduction
◆ Statement of significance
◆ Role and scope
◆ Definitions
◆ Literature review
◆ Theoretical basis
◆ Hypothesis
◆ Methodology
◆ Expectations and implications
◆ References (may come last)
◆ Appendix

Example 1. A Short Introduction in a Formal Proposal

What It Looks Like. In some fields and in certain instances the introduction can be used to explain the project's significance and to demonstrate the need to solve a problem. The following example (Figure 4.1) illustrates the means by which Kim Weatherly accomplished this.

Figure 4.1 Short Introduction from Proposal by Kim Weatherly

Patient Controlled Analgesia Versus Intramuscular Analgesia for Post Cesarean Pain Relief

Post-operative pain has long been recognized as a problem, especially during the first 24 hours after surgery. The problem is often related to the patient's inability to gain sufficient pain relief, which results in the patient being uncomfortable. This lack of pain relief may lead to an unsatisfactory surgical and hospital experience on the part of the patient. This can be especially detrimental for the new mother who has just undergone a cesarean section.

The hours immediately following birth are an important period of time for maternal and infant bonding. The mother has looked forward to the birth of her child for months and delivery by cesarean section and the resulting pain involved can interfere with this important bonding period. After a cesarean section the mother may experience intense pain, which may interfere with her ability to feed, care for, and bond with her new baby. Pain management is especially important for this group of patients because not only are they expected to recover from major abdominal surgery and resume caring for themselves, they are also expected to assume the tasks of caring for their newborn.

Traditionally intramuscular (IM) injections have been given in an attempt to control post-operative pain. With this method the patient requests pain medication that is given every three to four hours by the nursing staff on an as-needed basis. This often results in a delay between request and administration (Bennett, Batenhorst, Bivins, Bell, Graves, Foster, Wright, & Griffin, 1982). Therefore by the time the patient actually receives the injection, the pain is so severe that adequate pain relief is not obtained. Studies show that when acute post-operative pain is poorly managed it can have harmful psychological and physiological effects (Duggleby, 1994).

The traditional post-operative regiment of administering IM analgesics frequently fails to provide adequate pain relief: "Much of the inadequacy of the IM dosing regimen is due to the unpredictable and uneven absorption rate and the individual pain intensity and tolerance" (Jackson, 1989, p. 42). The inadequacies of IM narcotic administration for post-operative analgesia have led to the development of alternative techniques. Recent studies have demonstrated that patient-controlled analgesia (PCA) is an effective means of narcotic administration and pain control post-operatively (Rayburn, et al., 1988). The purpose of this study is to examine the effectiveness of intravenous (IV) PCA versus IM analgesia in the first 24 hours post cesarean.

How It Works. To accomplish its goal, the introduction must present some important information so that the reader knows the significance of this project. The danger when writing the introduction is that too much data can be presented, resulting in writing some of the literature review before the literature review has begun. Nevertheless, the writer must often draw upon the literature explored in order to convince the audience of the utility of the study. Here, in this introduction, it is vital that the reader understand the limitations of traditional procedure and the need for an alternative.

Example 2. Defining Terms: A Section from a Formal Proposal

What It Looks Like. In this portion of the document, Kim must explain the meaning of key phrases. I have reproduced three of her definitions (see Figure 4.2).

How It Works. Kim keeps the definitions simple, describing processes that will be referenced repeatedly in the proposal. Many times, elaborate definitions of terms are not necessary. Tables of definitions or abbreviations will appear instead.

Example 3. Statement of Purpose of the Study

What It Looks Like. Kim needed a concise statement of purpose to focus her proposal, as shown in Figure 4.3.

Figure 4.2 Defining Terms: From Proposal by Kim Weatherly

Definition of Terms

Patient-controlled analgesia: (PCA) is an electronically controlled infusion pump that allows the patient to receive a pre-set amount of narcotic by pushing a button.

Intramuscular analgesia: injection of a drug into a muscle.

Cesarean section: incision through the abdominal and uterine walls for delivery of a fetus.

Figure 4.3 Purpose Statement from Proposal by Kim Weatherly

Purpose of the Study

This study will evaluate the effectiveness of intravenous patient-controlled analgesia versus intramuscular analgesia in the first 24 hours post cesarean.

How It Works. The statement of purpose corresponds to a "thesis statement" in a conventional academic paper. It must be absolutely clear to the reader.

Example 4. A Review of Literature Embedded in a Formal Proposal

What It Looks Like. One informational feature of proposals for research projects and of certain institutional or corporate proposals is the review of literature, a section that surveys what is known and discussed concerning your subject area so that the reader understands the context in which your intended work is situated, and so that you yourself can demonstrate your knowledge of the field. Corporate and institutional use of this section is minimal and sparing—sometimes confined to remarks in an appendix, often supported by graphical material. Academic use of this component is extensive, as the reviewer of the prospectus must be convinced that the researcher knows the field and is qualified to synthesize a judgment about it. Figure 4.4 presents part of a literature review in Kim's formal research proposal.

Figure 4.4 Literature Review: From Proposal by Kim Weatherly

Literature Review

Historically, pain has been a concept that has been difficult to define. All definitions suggest that pain is a complex, unpleasant phenomenon that is uniquely experienced by each individual and cannot be adequately defined, identified, or measured. McCaffery defines pain as "whatever the experiencing person says it is, existing whenever he says it does" (Leo and Huether, 1988, p. 423).

Nurses play an important role in the management of post-operative pain. They are responsible for assessing the patient's pain level, administering the pain medication and evaluating its effectiveness, and assessing the need for further intervention. The goal in managing women who have undergone a cesarean section is to promote comfort and relieve pain, while at the same time leaving the new mother alert and able to care for herself and her infant (Olden, Jordan, Sakima, & Grass, 1995). Over the last 20 years a new method of pain relief has been developed in an attempt to provide post-operative patients with better pain relief with decreased sedation. Patient-controlled analgesia has been used with increasing frequency for many types of post-surgical patients.

Research into the use of PCA began in the 1960s with Sechzer doing the pioneer work. He evaluated the analgesic response to small doses of IV narcotics that were given on demand by a nurse observer. His conclusion was that demand analgesia provided greater pain relief with smaller drug doses than those used with traditional IM analgesia. Concurrently, similar research was being conducted that allowed patients to self-administer small IV doses of narcotics. It was this research that lead to the development of

---

(continued on next page)

Figure 4.4   *(continued)*

the 'Cardiff palliator,' the first commercially available PCA, developed at the Welsh National School of Medicine and produced in England (Collier, 1990). PCA has since evolved as a means to provide patients with adequate pain relief, maintain a more steady plasma level of narcotics, and give the patients greater flexibility in managing their own pain. PCA is a well-tolerated and effective method of pain control, especially in the post-operative period (Bucknell & Sikorski, 1989; Eisencah, Grice, & Dewan, 1988; Collier, 1990).

PCA is an electronically controlled infusion pump that allows patients to self-administer pain medication by pressing a hand-held button that is connected to the machine. Unlike earlier machines that had no control over how much medication the patient received or at what rate, newer machines can be programmed to deliver a specified amount of medication at predetermined intervals. The physician orders the type and dosage of medication. The nurse then sets up the PCA, programs the settings, and instructs the patient on its use. Medication can be administered either continuously, on patient demand, or both through an indwelling intravenous catheter. The dosage of medication and lock out intervals are set and then the machine is locked to prevent tampering. The lock out interval is used to prevent the patient from administering another dose of medication before the first dose has reached its peak and to prevent overmedication (Collier, 1990; Duggleby, 1994; Bennett, et al., 1982; Jackson, 1989).

There have been a number of medical and nursing studies that have compared the use of IV-PCA and nurse administered IM injections (Rayburn, et al., 1988; Jackson, 1989; Bucknell & Sikorski, 1989; Duggleby, 1994). The results of these studies vary, from showing improved pain relief with less medication, to providing improved pain relief with more medication, to demonstrating no difference between either method . . .

Rayburn, Geranis, Ramadei, Woods, and Ptil (1988) conducted a study that included 130 women who were delivered by cesarean section and randomly assigned to receive meperidine by IV-PCA or IM injection. Their conclusions were that PCA gave more adequate pain relief in higher but still safe doses and with less overall sedation than IM analgesia. They reported that the patient-controlled analgesia method provided less sedation and more immediate pain relief without the need for painful injections and was therefore a safe and effective means of satisfying individual patient analgesia needs after cesarean section. Bucknell and Sikorski's (1989) study of 40 women who had undergone cesarean section reached similar conclusions.

Similar findings have appeared in other studies with patients undergoing various types of surgical procedures. In 1989 Jackson published a study which compared patient-controlled analgesia with IM analgesia in 328 patients who had undergone elective cholecystectomy and abdominal hysterectomy. These patients were also randomly assigned to receive either IV patient-controlled analgesia or IM analgesia. These results demonstrated that PCA therapy provides for individualized dosage titration and therefore more optimal pain management. Their conclusions were similar to previous studies that found that PCA was a safe and effective method for postoperative pain management (Jackson, 1989).

How It Works. Notice that Kim uses some topic sentences including several sources to counterbalance the inevitably serial aspect of citation of research in which one study follows another one in time, and that is followed by another. In fact, because of the chronological nature of her topic, Kim informs the reader that there is an historical time line of research so that the reader expects this aspect to be apparent in the review. Balancing chronology against a conceptual organization can be difficult when constructing this section of your document.

Because such reviews are conceptual, not simply accumulative, the writer should not fill several pages with consecutive summaries of separate sources. That would merely result in a review that does not proceed by concept, but merely lists authors in what appears to be an arbitrary order. The reader should not receive the impression that the author has been stringing separate but unrelated summaries together. Rather, the review of literature must organize each sequence of paragraphs under a heading or topic sentence expressing a central idea. Remember, it is best to summarize the concept, citing several authors while doing so, rather than to summarize each author in a stand-alone fashion. The reader should not have to seek the concept supported by citations of authors. The reader should feel that the work of one cited author relates to that of another.

These literature reviews are especially challenging to write if the topic is new and the field has little or no precedent. But the inverse can also be trying for the writer who discovers an almost-infinite amount of material to sift through and ponder.

Example 5. Research Design: Sample, Instruments, Data Collection

What They Look Like. I've included these features of Kim's proposal to illustrate the logical flow of information presented in such a document (see Figure 4.5).

Figure 4.5 Details from Proposal by Kim Weatherly

Research Design

This is a descriptive, convenience sample. Participants will be chosen because of their availability and willingness to participate in the study.

The Sample

The population of this study will be women undergoing a cesarean section at participating hospitals during the study period. Participants will be

---

(continued on next page)

Figure 4.5 *(continued)*

assigned to either the IV-PCA or IM group based on their physicians' preference for prescribing post cesarean pain relief. Participants will be free of
accompanying serious medical illness, hepatic or renal disease, or history of
alcohol or drug abuse.

Instruments

An 11-point numerical rating scale (NRS) will be used to evaluate the
patient's pain. The numerical rating scale is a numbered line ranging from
0–10, with 0 being 'no pain' and 10 being 'worst possible pain' (Wong &
Waley, 1990). At the completion of the 24 hours the participants will be
asked to rate their satisfaction with their post-operative analgesia using the
same scale, with '0' representing 'completely unsatisfied' and '10' representing
'completely satisfied.'

Numerical rating scale for pain

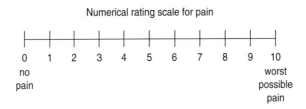

Numerical rating scale for satisfied

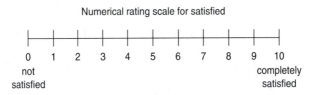

Method of Data Collection

Data collection will begin with initiation of either PCA or IM analgesia
post-operatively. A numerical rating scale (NRS) to assess pain will be completed on initiation of post-operative medication and continue every four hours
for 24 hours. Nurses at participating hospitals will be trained in data collection
by the nurse researcher to assure that all data is collected in the same manner. All patients who are scheduled for cesarean as well as laboring patients,
because they have the potential for a cesarean delivery, will be approached
prior to delivery to obtain consent. Upon enrollment into the study, nurses
will discuss with the participant the type of analgesia that her physician uses
and explain the PCA if indicated. Participants will be assured of confidentiality
and be aware that they can withdraw from the study at any time.

How These Work. These sections do not end the proposal, but are followed first by a statement on the protection of human subjects and institutional approval of the project, then by a list of references in APA style.

Further Comments on This Proposal
The actual order of all parts in Kim's proposal is this:

◆ Cover page
◆ Introduction
◆ Literature review
◆ Purpose of the study
◆ Definition of terms
◆ Research design
    Sample
    Instruments
    Method of data collection
    Reliability and validity
    Data analysis
◆ Protection of human subjects
◆ References

I was curious to know what Kim thought about this form of writing and whether it had helped her in other ways since school, so I asked Kim some questions.

Q: What do you do?
A: I'm an 11-7 charge nurse, obstetrics, labor and delivery, newborn nursery.

Q: Has knowledge of this technique and this style of writing helped you?
A: I now know how to conduct research and write proposals.

Q: What part of the document was easiest?
A: The literature review was fairly easy to do because of the Internet.

Q: And what posed the greatest challenge to you?
A: Finding a tool and the validity and reliability to back it up.

Q: Are there other aspects of your project you'd like to discuss, particularly as they relate to the proposal?
A: Data collection changed in the final study from what it was in the proposal. All participants were randomly assigned to

one of the two treatment groups—either IV-PCA or IM
based on their hospital admission number. Those in the IM
group were given Demerol every four hours on an as-
needed basis and those in the PCA group were given
Demerol via PCA.

Q: Do you now use proposal writing at work?

A: At my current position I personally do not use it, but in
the next two to three months I will be going to work as
a family nurse practitioner in a clinic where I'm sure I
will use it to help write grants.

Example 6. An Article-Length Proposal

Context.   I asked sociology professor Janice Clifford Wittekind to supply a
proposal for this text—one that is about the size of an article. This proposal
is twenty-six pages in typescript and adheres to the APA style current at the
time of its writing. I also interviewed her, asking Janice to explain a number
of issues about such documents that can be puzzling for students. For example, why write a proposal in the first place? Of what good is it? Portions
of this interview follow.

Q: What do you do?

A: As a professor of sociology and criminal justice, I assist
students in gaining a more comprehensive understanding of society. One way to facilitate this is by challenging students' abilities to think critically about the
world they live in through the development of a research project.

Q: How do you use proposal writing at work?

A: I teach courses in research methods. During the
semester, the students work through the process of
developing a full-scale proposal including significance of
the study, hypotheses, definition of variables, methodology and expected findings. Proposal writing is also
used when I conduct research. It is critical to have a
well-thought-out agenda, and one way to ensure this is
to develop a proposal for the project.

Q: How has knowledge of this style and technique helped
you?

A: It has assisted me in writing papers in graduate school
and when preparing documents for submission for
publication in academic literature and presentations at
professional conferences.

Q: What is your greatest challenge when constructing a proposal or similar document?

A: My greatest challenge is to logically set up the framework for the proposed research and to be aware of any pitfalls in the research design. Another challenge is to be consistent in terms of formatting.

Q: What tips can you offer someone learning to write in this style?

A: It takes several tries to become familiar with APA formatting. It is important that the student be aware that mastery of this style takes time. One thing that really helps in the learning process is to own one's own copy of the APA style manual and to use it as one is writing the document. I have a copy on my shelf in the office and it serves as my reference guide for writing.

Q: Describe the context and mission of the particular item you have contributed.

A: The proposal that I have contributed derived from part of my dissertation work in graduate school. The content of the proposal provides students with a social science model of proposal writing.

Figure 4.6 reproduces Janice's proposal. Short annotations follow.

Figure 4.6   Proposal by Janice Clifford Wittekind, Copyright (c) 1999
Janice Clifford Wittekind and Used by Permission

Unwed Mother   1

Does Growing Up in a Single-Mother Family Contribute
to a Female's Chances of Becoming an Unwed Mother?

Dr. Janice Clifford Wittekind
The College of West Virginia

Proposal by Janice Clifford Wittekind, Copyright (c) 1999 Janice Clifford Wittekind and Used by Permission

Figure 4.6 *(continued)*

Abstract

The number of female-headed, single-parent homes has increased over the past decade. Although numerous studies have examined the effects of single-mother families on adolescents, few studies have assessed the long-term consequences on adult functioning of being raised in a single-mother home in childhood. This study guided by social learning theory and role theory examines whether childhood family structure affects adult children's intimate relationships. Using data from the National Survey of Families and Households (n=13,008), this study will examine whether the experience of living in a single-mother family in childhood affects the likelihood of becoming an unwed mother. The effects of being raised in a single-parent family will be accessed through the mediating variables of psychological distress, alcohol abuse and courtship values.

Proposal by Janice Clifford Wittekind, Copyright (c) 1999 Janice Clifford Wittekind and Used by Permission

(continued on next page)

Figure 4.6 *(continued)*

Table of Contents

Proposal by Janice Clifford Wittekind, Copyright (c) 1999 Janice Clifford Wittekind and Used by Permission

Figure 4.6 *(continued)*

I.   Introduction

Over one half of the children born today will spend some portion of their childhood in a single-parent home (Lamanna and Reidmann, 1994; Ellwood, 1987). The number of children under eighteen years of age who lived in a single-parent family has risen substantially over the past several decades: 1970 - 8.2 million, 1980 - 12.5 million to 1990 - 15.9 million, 25% - 1990 (Schmittroth, 1992: 823).

Previous research has focused on some of the ways in which single-mother families affect child and adolescent development. In general, reviews of that research (Amato and Keith, 1991; Wallerstein and Blakeslee, 1989; Weitzman, 1985) indicate that children growing up in single-mother homes score somewhat lower in social skills, score higher in problem behavior and have difficulty in academic functioning. While findings have indicated that being raised in a single-mother family has effects when the child is young and still part of the family (Demo and Acock, 1988; Wallerstein, Corbin and Lewis, 1988; Hetherington and Camara, 1988; Krantz, 1988), there has been little research examining the long-term effects in adulthood.

Research examining the consequences of being raised in a single-mother family found that these adults are distinguished by an increase in levels of psychological distress, marriage at an earlier age, more premarital births, lower educational attainment levels and a greater likelihood of divorce (Wallerstein, Corbin and Lewis, 1988; Wallerstein and Blakeslee, 1989; Beal and Hochman, 1991; McLanahan, 1991; McLanahan and Bumpass, 1988; Wallerstein and Kelly, 1974).

Proposal by Janice Clifford Wittekind, Copyright (c) 1999 Janice Clifford Wittekind and Used by Permission

(continued on next page)

Figure 4.6 *(continued)*

However, while findings have indicated that being raised in a single-mother family has effects when the child is young and still part of the family, there has been little research examining the long-term effects in adulthood. This paper, using data from the National Survey of Families and Households, examines whether the experience of living in a single mother family in childhood affects the stability of intimate relationships in adulthood.

Previous research on the topic has been limited by the use of cross-sectional data and retrospective accounts of respondents' marital histories and socialization. In this analysis childhood family background data is used to predict the odds of adult relationship instability. The proposed research examines whether growing up in a single-mother home affects current behavior in adulthood through intervening variables that in turn affect relationship stability. The three variables are: psychological distress, alcohol abuse, and attitudes toward cohabitation.

This study examines whether the experience of living in a single-mother family in childhood affects daughters' chances of becoming unwed mothers. Based on a review of both theory and research, it is argued that growing up in a single-mother family: a) provides a model of single adulthood; b) deprives children of a model of intimate heterosexual interaction; c) deprives children of gender role models. These experiences influence the formation of an individual's personality characteristics and values, which in turn affects stability of intimate relationships. These questions will be addressed through the application of social learning theory and role theory.

Proposal by Janice Clifford Wittekind, Copyright (c) 1999 Janice Clifford Wittekind and Used by Permission

Figure 4.6 *(continued)*

II. Significance of the Topic

The effects of single-mother families on children are important to study because the divorce rate has steadily increased over the past decades. In 1995, 16,477,000 children under the age of 18 were living in mother-only families. Although single-father families have risen in the 1990s, they still represent a small minority of the existing family structures. The following figures relate to the percentage of children under age 18 in the population who experience divorce: 1970 - 12.5%; 1980 - 17.3%; 1985 - 17.3%; 1988 - 16.4%; 1990 - 16.8% (U.S. Bureau of the Census, 1996:105).

Despite all the past research, little is known about the mechanisms by which growing up in a single-mother family leads to unwed motherhood. This study explores the mechanisms by which single-mother families have long-term effects on adult relationships. The questions it raises are: "Does growing up in childhood in a single-mother family lead to personality and behavioral characteristic deficits that make instability in intimate relationships in adulthood more likely?" Or, "Is it the observance of model of a single-mother's liberal dating behaviors that contribute to children's non-conventional courtship values, resulting in increased relationship instability in adulthood which contributes to unwed motherhood?"

The information presented in the subsequent literature review provides a basis for linking the current work to previous research. In conducting research, it is important to examine what has previously been done on the topic. This prevents exact replication of studies and forces present research to examine the topic from a different perspective, in order to contribute to advancing knowledge in the field of Sociology of the Family.

Proposal by Janice Clifford Wittekind, Copyright (c) 1999 Janice Clifford Wittekind and Used by Permission

(continued on next page)

Figure 4.6 *(continued)*

III.   Literature Review

Few studies have examined the relationship between childhood family structure and adult relationship stability. The majority of past research has focused on the immediate or short-term consequences of divorce and the single-parent family on children. Studies (Wallerstein et al., 1974, 1988, 1989; Wallerstein and Kelly, 1980; McLanahan and Bumpass, 1986; 1988; Wu and Martinson, 1993; Hogan and Kitagawa, 1985) that do examine the long-term effects explain their presence though social, structural and situational variables. They do not address the role of personality, behavioral or value characteristics.

Children add a fundamental and important dimension to family life. Most people agree that the "ideal family for the job is a nuclear family, preferably with both male and female adults present to act as care givers and role models" (Jones, Tepperman, and Wilson, 1995:103). In 1990, there were about 63 million children under the age of 18 living in American households. Yet, twenty-eight percent lived with only one-parent, nearly always their mother (Jones, Tepperman, and Wilson, 1995:103).

The best evidence supporting the intergenerational replication of female household headship is presented by the work of Hogan and Kitagawa (1985) and Hogan (1985). Their results indicate that adolescent females from mother-headed families, compared to adolescent females from two-parent families, were more likely to engage in sexual activities and to give birth out-of-wedlock (Hogan and Kitagawa, 1985; Hogan 1985). Research that focused on the relationship between childhood family structure and adulthood outcomes in the late 1980s and early 1990s revealed that there are long-range consequences of being raised in a single-mother home. Studies found that these adult children have lower levels of educational attainment, have sexual

Proposal by Janice Clifford Wittekind, Copyright (c) 1999 Janice Clifford Wittekind and Used by Permission

Figure 4.6 *(continued)*

relations at an earlier age and are more likely to have pre-marital births (Astone and McLanahan, 1991; Bumpass and McLanahan, 1989; Krein and Beller, 1988; Hogan and Kitagawa, 1985; McLanahan and Bumpass, 1988).

Regardless of the process of how the single-mother family was formed, it generates more chronic and acute stressors. The experience of divorce imposes a specific stressor on adolescents in dealing with a loss of a parent. These outward signs of stress may appear through the presence of depression, representing internalization, or the externalization exhibited by problem behavior through delinquency or substance abuse.

It is stated that the responses to the stresses accompanying family disruption and turmoil inflict negative consequences on children and adolescents. Previous studies indicated that children residing in single-parent households experience a greater number of stressful events and have an increased sensitivity to them (Belle, 1984; McAdoo, 1986; Pearlin and Johnson, 1977; Weinraub and Wolfe, 1983). Findings consistently show that stressful events are always associated with negative behavioral outcomes.

Studies of stress, physical and psychological functioning among adolescents found a relationship between the presence of stress and acting-out behavior and the presence of depression (Felner, Stolberg and Cowen, 1975; Johnson and McCutcheon, 1980). Research conducted with adults presents additional support for the link between stressful life events and psychological problems, ranging from anxiety to depression and social maladjustment (Rabkin and Streuning, 1976; Thoits, 1983).

Proposal by Janice Clifford Wittekind, Copyright (c) 1999 Janice Clifford Wittekind and Used by Permission

(continued on next page)

Figure 4.6 *(continued)*

Conclusions

How does childhood family structure affect relationship stability and likelihood of unwed motherhood? One expected finding is that due to the increased number of stressors that children in mother-only families experience in childhood, they will carry forward into adulthood the deficits that they obtained in childhood. These are believed to be the outward signs in childhood: psychological distress and problem behavior that is exhibited by alcohol abuse.

It is argued that the short-term impact of being raised in a single-mother home has effects which may be influential later in life when adults are involved in intimate relationships. It is speculated that these long-term effects are carried over from childhood.

Proposal by Janice Clifford Wittekind, Copyright (c) 1999 Janice Clifford Wittekind and Used by Permission

Figure 4.6 *(continued)*

IV. Theoretical Perspective

Social Learning Theory (Modeling)

One approach useful in explaining the long-term effects of being raised in a single-mother family on the marital stability of adult children is the application of social learning theory. Social learning theorists believe that behavior is learned through the observation of models being reinforced or punished. Models provide others with styles of thought and conduct, and serve to strengthen or weaken inhibitions over behavior that the observers have previously learned (Bandura, 1971). This process is important, especially in interpersonal relationships with others, as the style of communicating and problem-solving are based on observed models of interaction (Bandura, 1977). In a single-mother family, children are deprived of models of marital communication and problem-solving. The lack of a male and a female model deprives children of opportunities to learn appropriate interaction skills that are necessary to sustain an intimate relationship.

This application of social learning theory hypothesizes that children mimic their parents' behavior. The traditional two-parent family structure (a mother and father) is believed to present children with a model of behavior and interpersonal communication, which include intimacy, problem-solving, conflict resolution, and heterosexual relations. In the case of single-mother families, children do not have an appropriate model to imitate, and thus lack the appropriate role model.

Through the experience of living in a single-mother family and the subsequent socialization process, children's chances of developing personality traits and behavior that interfere in intimate relationships in adulthood are greatly increased (Amato, 1996). In families that have discord or fail to exhibit affection between parents, children lack prolonged exposure to models

Proposal by Janice Clifford Wittekind, Copyright (c) 1999 Janice Clifford Wittekind and Used by Permission

(continued on next page)

Figure 4.6 *(continued)*

of successful dyadic interaction. Consequently, these children may not learn the necessary interpersonal skills (such as compromising and proper communicating) that promote mutually satisfying, long-term ties with others.

The experience of observing a single-mother's liberal dating relationship may translate into non-conventional attitudes towards dating and marriage in her children. Studies show that in adulthood, children of divorce and children raised in never-married mother homes are more likely to cohabit prior to marriage (Bumpass, Sweet and Cherlin, 1989; Furstenberg and Teitler, 1994; Thornton, 1991).

Gender Role Theory

Gender role theory applies to the process of learning how to feel, behave and view the world in a way similar to those who occupy the same gender role (Backman, 1964). It includes the learning of the attitudes, skills and techniques associated with that role. Roles consist of expectations for behavior characteristics of a recognized set of persons in a specific social position or status, which carry with them expectations for appropriate conduct (Turner, 1978). The underlying tenets of role theory are: 1) that it provides a way of linking individual behavior with larger social groups through the concepts of status or position in society; 2) persons enact roles, but positions underlying them are determined by society; 3) participation in many roles simultaneously is subject to cost and benefits; and 4) individuals are more or less suited to a role (Turner, 1978; Goode, 1960).

When examining the development of sex roles, role theory describes the process by which children learn their gender identity. It suggests that males and females learn how to be "male" and "female" by interaction with another same-sexed person, and by playing a complementary role

Proposal by Janice Clifford Wittekind, Copyright (c) 1999 Janice Clifford Wittekind and Used by Permission

Figure 4.6 *(continued)*

with someone of the opposite sex. This theory suggests that being raised in a single-parent home denies children a partner with whom to identify or rehearse. Females raised in a female-headed family lack an opposite-sexed partner with whom to rehearse a feminine role.

Conclusions

The long-term effects of single-mother families on adult relationships present themselves through their effects on psychological distress, problem behavior (drug, alcohol abuse or delinquency and crime) and scripts for courtship and marital relations. Single-mother homes expose children to stress and place them in situations where they do not have the appropriate skills to cope. This inability to handle stress adequately often leads to personality and behavioral deficits. The stress and deprivation in single-mother families lead to lingering psychological distress in females.

Single-mother families have direct and indirect effects on children's intimate relationships in adulthood. The direct effects are a consequence of modeling. The lack of role models to observe in a successful intimate relationship deprives children of scripts for appropriately dealing with the stresses and conflicts in marriage. Their inability to observe and rehearse appropriate "sexed-type" behavior also leaves them with scripts deficits for interpersonal communication, dating behavior and marital relationships.

The model of a single-mother family conveys non-conventional attitudes about courtship and cohabitation. It is argued that living in a single-mother family liberalizes children's attitudes toward divorce and increases the chance that they will exhibit deficits in interpersonal skills and behavior. As adults, these individuals are more likely to cohabit before marriage.

These findings suggest that parental marital status has a direct impact on children's interpersonal behaviors. More specifically, children raised in single-mother families are not exposed to models of dyadic behavior

Proposal by Janice Clifford Wittekind, Copyright (c) 1999 Janice Clifford Wittekind and Used by Permission

(continued on next page)

Figure 4.6 *(continued)*

and may not learn the skills and attitudes required that facilitate successful functioning in marital roles. These are plausible routes to explain the presence of long-term consequences. The long-term effects manifest in deficits in appropriate patterns for courtship behavior. These deficits do not emerge until adulthood when individuals are placed into a situation where their resources are drawn upon.

Individuals who uphold non-conventional courtship values have models for love relationships that represent their readiness to cohabit. Their experience in a single-mother family has transmitted non-conventional values to children regarding cohabitation, marriage, and single parenthood. Thus, one expects to find that children raised in single-mother home have a greater chance of cohabiting in adulthood. Cohabiting relationships can be unstable since they lack a marital commitment, and can consequently be an avenue for unwed motherhood.

Proposal by Janice Clifford Wittekind, Copyright (c) 1999 Janice Clifford Wittekind and Used by Permission

Figure 4.6   *(continued)*

V. Hypothesis

The goal of this study is to uncover the effects of growing up in a single-parent household on the perceived chance of daughters' experiencing unwed motherhood. In addition, the objective is to explore questions about the mechanisms through which childhood family structure generates long-term consequences in adulthood.

This study, guided by social learning theory and role theory, examines the relationship between being raised in a single-mother family and the stability of adult children's intimate relationships. It will identify three intervening variables through which being raised in a single-parent family during childhood influence the chance of unwed motherhood: psychological distress, alcohol abuse, and the transmission of non-conventional courtship values. Psychological distress is examined as it is theorized that mental functioning is affected by stressors that are inflicted by being raised in a single-parent family.

Alcohol abuse is incorporated as it is speculated that females exhibit acting-out behavior in response to stress, most notably represented through drinking habits.

Courtship values are of interest because being raised in a single-parent home often displays an approval of non-conventional courtship values (e.g., cohabitation and dating by mother). This would account for the replication of single-mother families by subsequent generations. The hypotheses are as follows:

H1)   Never married adult children raised in single-mother families have a greater chance of becoming unwed single-mothers.

H1a)   The effects of being raised in a single-mother family on a never-married female's chances of becoming an unwed mother are mediated through the intervening variables of psychological distress, alcohol abuse, and courtship values.

(Refer to Diagram #1)

Proposal by Janice Clifford Wittekind, Copyright (c) 1999 Janice Clifford Wittekind and Used by Permission

(continued on next page)

Figure 4.6   *(continued)*

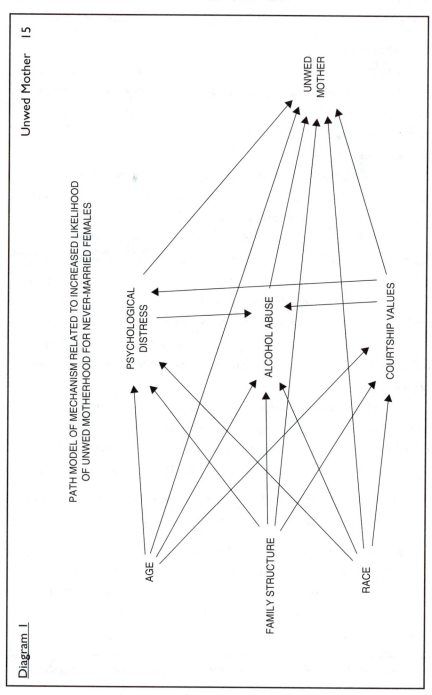

Proposal by Janice Clifford Wittekind, Copyright (c) 1999 Janice Clifford Wittekind and Used by Permission

Figure 4.6   *(continued)*

VI.   Methodology

Research Design

The type of research methodology to be performed is secondary data analysis of cross-sectional survey data from the National Survey of Families and Households (NSFH). The data were collected between 1987–1988, by the Center for Demography and Ecology at the University of Wisconsin-Madison. The two different sections of the data set included in this study were:

1. Main interview schedule: the interview schedule administered to the primary respondent by the interviewer. However, several portions of the main interview were self-administered to facilitate the collection of sensitive information and to ease the flow of the interview.

2. The self-administered questionnaire: the self-administered form filled out by the primary respondent at various points during the course of the interview.

Family structure can be defined in many ways. Due to the design of the data set, it is not possible to determine the specific type of single-parent family structure the respondent grew up in. The data set to be used for this study contains information on adult children who grew up in all possible formations of single-parent families in childhood—separation, divorce, widowhood, adoption or unwed motherhood.

Sampling Issues

The National Survey of Families and Households contains interviews with a probability sample of 13,008 respondents. The survey portion of the data set includes a main sample of 9,643 respondents who represent the non-institutionalized population age 19 and older, who resided in the United States in 1987 and 1988. In addition, several population groups were

Proposal by Janice Clifford Wittekind, Copyright (c) 1999 Janice Clifford Wittekind and Used by Permission

(continued on next page)

Figure 4.6   *(continued)*

over-sampled: minority groups (African Americans, Puerto Ricans and Chicanos), single-parents, persons with step children, cohabiting persons and persons who recently married (Center for Demography and Ecology, 1990).

Measurement Issues

The proposed variables believed to affect the respondent's likelihood of relationship instability leading to unwed motherhood are measured using variables which determine a respondent's age, race, childhood family structure, psychological state, presence of alcohol abuse and courtship values (conventional or non-conventional). The intervening variables include measures of psychological distress, alcohol abuse and non-conventional courtship values that represent cohabitation history or attitudes/opinions towards cohabitation.

The dependent variable indicates whether or not the never-married female is an unwed mother. A respondent is considered to be an unwed mother if her marital status is never-married and she reported parenting one or more children.

Childhood family structure is the preceding exogenous independent variable of most significance. The definition of family structure accounts for variations—two biological parents or a single-parent household. The independent variable is the type of family structure the respondent lived in up to age 19. The question asked, "Did you live with both your biological parents from the time you were born until age 19?" The possible responses are: 0—no and 1—yes.

A respondent's overall psychological state is determined through the construction of a multiple item scale based on the question in the data set which asked how many days per week did the respondent feel: 1) depressed, 2) fearful, and 3) sad. These will be summed together to create a scale measuring level of psychological distress. The scale range is 0 – 21 days.

Proposal by Janice Clifford Wittekind, Copyright (c) 1999 Janice Clifford Wittekind and Used by Permission

Figure 4.6 *(continued)*

The values for the scale are: 0—no psychological distress present to 21—high level of psychological distress.

An assessment of whether or not the respondent has an alcohol problem is determined by a self-report to a series of questions. The first question asks, "Does anyone living [in the respondent's home] have a drug or alcohol problem?" The second question asks, "Who living here [in the respondent's home] has a problem of drinking too much alcohol?" The responses are: 0—other person and 1—respondent. To determine whether or not the respondent abuses alcohol, a new variable will be created indicating that if the respondent is the identified person in the house with a drinking problem, then alcohol abuse exists.

An indication of the respondent's conformity or non-conformity to traditional societal courtship values was identified. For never-married females, an assessment of their values is determined by their response to the statement, "Importance of the reason why they would not cohabit with a member of the opposite sex because it is morally wrong." The range of responses is: 1—"very important" to 7—"not important at all." A low score indicates conventional courtship values and a high score indicates non-conventional courtship values.

Age and race are entered into the proposed model as exogenous control variables, as stated previously, because they may be independently related to relationship and/or marital instability. The significance of their inclusion was discussed earlier under the hypothesis section.

Data Analysis

A respondent's likelihood of unwed motherhood is predicted to be influenced by many factors, both internal and external to the individual. In an attempt to explain the hypothesized mechanism(s), which affect the stability of intimate relations, regression and path modeling will be conducted.

Proposal by Janice Clifford Wittekind, Copyright (c) 1999 Janice Clifford Wittekind and Used by Permission

(continued on next page)

Figure 4.6 *(continued)*

Through these statistical applications, the direct effects of family structure on the likelihood of unwed motherhood for a never-married female can be determined. Further analysis examines the direct effect of family structure and its indirect effects on the dependent variable through independent intervening variables of psychological distress, alcohol abuse, and courtship values.

Proposal by Janice Clifford Wittekind, Copyright (c) 1999 Janice Clifford Wittekind and Used by Permission

Figure 4.6 *(continued)*

VII.   Expected Findings

It is expected that after conducting statistical analysis on the data, evidence will emerge supporting the proposed relationship between childhood family structure and unwed motherhood. More specifically, being raised in a single-parent family structure will prove to have a significant effect on the chances of a never-married female becoming an unwed mother.

It is the lack of a stable model of successful dyadic interaction in the home that places children at a greater risk for repeating the pattern of single parenting. Further it is expected that the intervening variables of psychological distress, alcohol abuse, and value structure also influence the chances of becoming an unwed mother either directly or as mediating variables for family structure.

From the findings, support is expected to be found for the application of social learning theory and gender role theory as plausible explanations for the intergenerational repetition of female one-parent family structures.

Proposal by Janice Clifford Wittekind, Copyright (c) 1999 Janice Clifford Wittekind and Used by Permission

(continued on next page)

Figure 4.6 *(continued)*

VIII.   Implications

The evidence from this study will not account for all of the possible contributory factors that may affect relationship stability. However, it will contribute to our knowledge of the significance that childhood family structure, psychological distress, alcohol abuse, and courtship play. Additionally, it will advance the field of Sociology of the Family, in that it examines the significance of childhood family structure on adult intimate relationships utilizing quantitative data and applying historical theories to contemporary issues.

This study lays the foundation for future research to examine the ways in which single-mother families may be restructured to help reduce children's deficits in interpersonal, problem-solving, and communication skills. If children could master competence in these important areas, they would possess the necessary skills to maintain stable adult intimate relationships. The observance of a powerful female figure, the single mother, contributes to female children's acceptability of being independent from males.

Implications from this study go well beyond the academic realm. Conclusive findings will assist social service providers in acknowledging and addressing personal and behavioral characteristics as factors contributing to relationship instability and creation of the single-parent family.

Proposal by Janice Clifford Wittekind, Copyright (c) 1999 Janice Clifford Wittekind and Used by Permission

Figure 4.6 *(continued)*

Unwed Mother   22

References

Amato, P.R. (1996). Explaining the intergenerational transmission of divorce. Journal of Marriage and the Family, 58, 628-640.

Amato, P.R. and K.B. (1991). Parental divorce and the well-being of children: A meta analysis. Psychological Bulletin, 110 (1), 26-46.

Astone, N.M. and McLanahan, S. (1991). Family structure, parental practices and high school completion. American Sociological Review, 56, 309-320.

Bandura, A. (1977). Social learning theory. Englewood Cliffs, New Jersey: Prentice-Hall.

Bandura, A. (1971). Vicarious and self-reinforcement processes. In R.Glazer (Ed.), The nature of reinforcement. New York: Academic Press.

Beal, E.W. and Hochman, G. (1991). Adult children of divorce. NewYork: Delacorte Press.

Belle, D. (1984). Inequality and mental health: Low income and minority women. In L.Walker (Ed.), Women and mental health policy. Beverly Hills: Sage Publications.

Bumpass, L.L. and McLanahan, S. (1989). Unmarried motherhood: Recent trends, composition, and black-white differences. Demography, 26, 279-86.

Bumpass, L.L., Martin, T.C., and Sweet, J.A. (1991). The impact of family background on early marital factors and marital disruption. Journal of Family Issues, 12, 22-24.

Center for Demography and Ecology. (1990). National survey of families and households: Codebook. University of Wisconsin-Madison.

Proposal by Janice Clifford Wittekind, Copyright (c) 1999 Janice Clifford Wittekind and Used by Permission

(continued on next page)

Figure 4.6 *(continued)*

Demo, D. and Acock, A. (1988). The impact of divorce on children. Journal of Marriage and the Family, 50, 619-648.

Ellwood, D.T. (1987). Divide and conquer: Responsible security for America's poor. Occasional Paper No.1. New York: Ford Foundation Project on Social Welfare and American Future.

Felner, R.D., Stolberg, A.L. and Cowan, E.L. (1975). Crisis events and school mental health referral patterns of young children. Journal of Consulting and Clinical Psychology, 43, 305-310.

Furstenberg, F.F. Jr. and Teitler, J.O. (1994). Reconsidering the effects of marital disruption: What happens to children of divorce in early adulthood. Journal of Marriage and the Family, 49, 811-825.

Goode, W. (1960). A theory of role strain. American Sociological Review, 25, 483-496.

Hetherington, E.M. and Camara, K. (1988). The effect of family dissolution and reconstitution on children. In N. Glenn and M. Coleman (Eds.), Family relations: A reader. New York: Dorsey Press.

Hogan, D.P. (1985). Structural and normative factors in single parenthood among black adolescents. Unpublished paper. Department of Sociology, University of Chicago.

Hogan, D.P. and Kitagawa, E.M. (1985). The impact of social status, family structure and the neighborhood on the fertility of black adolescents. American Journal of Sociology, 90, 825-855.

Proposal by Janice Clifford Wittekind, Copyright (c) 1999 Janice Clifford Wittekind and Used by Permission

Figure 4.6   *(continued)*

Unwed Mother   24

Johnson, J. and McCutcheon, S. (1980). Assessing life stress in older children and adolescents: Preliminary findings with the life events checklist. In I. Sarason and C. Spielberger (Eds.), Stress and anxiety. (v.6). New York: John Wiley.

Jones, C., Tepperman, L., and Wilson, S. (1995). The future of the family. Upper Saddle, New Jersey: Prentice Hall, Inc.

Krantz, S.E. (1989). The impact of divorce on children. In A.S. Skolnick and J.H. Skolnick (Eds.), Family in transition: Rethinking marriage, sexuality, child rearing and family organization. Glenview, Illinois: Scott, Foreman and Company.

Krein, S.F., and Beller, A. (1988). Educational attainment of children from single-parent families: Differences by exposure, gender and race. Demography, 25, 221-34.

Lamanna, M.A. and Reidmann, A. (1994). Marriage and families: Making choices and facing change. Belmont, California: Wadsworth.

McAdoo, Harriette Pipes. (1986). Strategies used by black single mothers against stress. In M. Simms and J. Malveaux (Eds.), Slipping through the cracks: The status of black women. New Brunswick, New Jersey: Transaction Books.

McLanahan, S. (1991). The long-term effects of family dissolution. In B. Christensen (Ed.), When families fail: The social costs. New York: University Press of America for The Rockford Institute.

McLanahan, S. and Bumpass, L. (1988). Intergenerational consequences of family disruption. Journal of Marriage and the Family, 94, 130-52.

McLanahan, S. and Bumpass, L. (1986). Intergenerational consequences of family disruption. Paper presented at the annual meeting of the Population Association of America.

Proposal by Janice Clifford Wittekind, Copyright (c) 1999 Janice Clifford Wittekind and Used by Permission

(continued on next page)

Figure 4.6   *(continued)*

Unwed Mother   25

Pearlin, L. and Johnson, J. (1977). Marital status, life-strains and depression. <u>American Sociological Review, 42</u>, 704-715.

Rabkin, J. and Struening, E. (1976). Life events, stress and illness. <u>Science, 194,</u> 1013-1020.

Schmittroth, L. (Ed.). (1994). <u>Statistical record of children</u>. Detroit: Gale Research, Inc.

Thoits, P.A. (1983). Dimensions of life events that influence psychological distress: An evaluation and synthesis of the literature. In H.B. Kaplin (Ed.), <u>Psychological stress.</u> New York: Academic Press.

Thornton, A. (1991). Influences of marital history of parents on the marital and cohabitational experiences of children. <u>American Journal of Sociology, 90,</u> 868-894.

Turner, R.H. (1978). A Theory of Role Strain. <u>American Journal of Sociology, 84,</u> 1-23.

U.S. Bureau of the Census. (1996). <u>Statistical abstracts of the United States: 1996.</u> (116th Edition) Washington, D.C.: Reference Press.

Wallerstein, J.S. and Blakeslee, S. (1989). <u>Second chances: Women, men and children: A decade after divorce</u>. New York: Ticknor & Fields.

Wallerstein, J.S., Corbin, S.B., and Lewis, J.M. (1988). Children of divorce: A 10-year study. In E.M. Hetherington and J.D. Arasteh (Eds.), <u>Impact of divorce, single-parenting and step-parenting on children.</u> Hillsdale, New York: Lawrence Earlbaum Associates.

Proposal by Janice Clifford Wittekind, Copyright (c) 1999 Janice Clifford Wittekind and Used by Permission

Figure 4.6 *(continued)*

Wallerstein, J.S. and Kelly, J.B. (1974). The effects of parental divorce: The adolescent experience. In J. Anthony and C. Kouprtnik (Eds.), <u>The child and his family: children at psychiatric risk.</u> New York: Wiley.

Wallerstein, J.S. and Kelly, J.B. (1980). <u>Surviving the breakup: How children and parents cope with divorce.</u> New York: Basic Books.

Weinraub, M. and Wolfe, B. (1983). The effects of stress and social supports on mother-child interactions in single-and two parent families. <u>Child Development, 54,</u> 1297-1311.

Wu, L., and Martinson, B. (1993). Family structure and the risk of a premarital birth. <u>American Sociological Review, 85,</u> 210-232.

Proposal by Janice Clifford Wittekind, Copyright (c) 1999 Janice Clifford Wittekind and Used by Permission

Notes Regarding This Proposal

Cover sheet. The cover sheet, as well as the rest of the document, adheres to a format expected by the discourse community reading this proposal. It has a running head, "Unwed Mother," that continues throughout the document. The choice of title is important. Janice keeps the phrasing simple and refrains from using a colon-based title, the bane of much academic writing. For instance, she does not entitle the work "Growing Up: A Single-Mother Family and a Female's Chances of Becoming an Unwed Mother," phrasing that is less natural and inherently pompous.

Abstract. This abstract is short—127 words if compound expressions are counted as two words. The abstract nevertheless provides the information needed by the reader to understand what the proposal is about. Remember that readers of proposals must understand what you are doing if they read only this initial summary.

Table of contents. Janice uses the contemporary style that no longer requires linking topic to page number with a string of periods like this:

Topic 5

Note that Janice's table of contents represents her entire argument—her structure and content of presentation—at a glance.

Introduction. This section begins with startling facts that indicate a problem, thereby engaging the reader's attention. Next, the introduction summarizes the scope of previous research, indicating a need for further inquiry and refinement that the proposed project can supply. Janice is careful to provide information that motivates us to read the rest of the document, but does not review the literature here.

Significance of the topic. This "department" in the document reinforces the need for the proposed study, affirming that the topic is socially relevant and that more needs to be known about it. It contains ideas that the later "Implications" section revisits after we have been made familiar with the literature and theory related to the topic. This "Significance" unit acts as a transition from the introduction to the literature review that follows it.

Literature review. The literature review explores observed effects and findings. Such a survey of previous research expands our perception of the problem's context and potential causes, but—because many questions remain—also reinforces the need for the project proposed. Janice is careful to use topic sentences that provide the reader with conceptual understanding rather than organizing paragraphs around enumerations of sources. The concluding segment functions not merely as a summary but also as a transition to the section on theoretical perspective. Showing the connections between the ideas rather than listing the ideas is the mark of a proper review.

Theoretical perspective. This portion explains what has been proposed as a theoretical context for the observed effects.

Hypothesis. The study's hypothesis includes a corollary, that the impact of being raised in a single-mother family is mediated through the variables described and also depicted on the chart. Hypotheses may be qualified in a variety of ways, and explanatory or stipulating discussions may appear in such a section. In some cases such sections receive separate headings.

Methodology. This unit covers research design, sampling, measurement, and the analysis of data. It describes collection and problems with achieving representational data. In discussing the procedure of interpretation, the proposal states how the classification and assessment of the data will occur.

Expected findings. This segment affirms that the process of data analysis will produce support for the conjecture of the proposal and will lead to correlative expectations, including the reinforcing of the theoretical explanations previously discussed.

Implications. This part begins by acknowledging that the study is limited in its inclusiveness and cannot incorporate all aspects. Then the section reaffirms the value of the study and its implications for scholars and social services. This section reminds readers of the significance of the study, now clarified because of our exposure to the literature, theory, and methodology.

References. The reference section is written in APA format current at the time of the document's construction. All citation systems are inherently unstable, however, and only the most recent version of a documentation guide should be consulted.

Retrospection. I asked Janice two further questions, retrospective ones about the positives and the negatives involved in constructing such a project. Her answers follow.

Q: In this document, what was easy to do and why?

A: The part that I enjoyed writing the most was the literature review. When conducting a literature review you become more aware of and familiar with the body of existing knowledge about the topic you propose to study. Writing a literature review is challenging, especially when there is a large amount of information available, as a determination needs to be made about what is most applicable to your current study.

Q: What in the document was difficult, and why?

A: The most difficult part for me is usually the methodology. There are so many ways to design a research project and options for data collection and analysis; it takes a great deal of time to think through. It is necessary to determine what type of methodology will be most applicable to access the outcome goals.

Moral: The proposal is your plan, but in order to succeed, this document itself must have a plan.

Example 7. Workplace Proposal Draft Expanding into a Thesis Proposal

Context. Paula Fields wanted to perform a research project involving limiting the spread of hospital-originated infections through incentives for staff hygiene. She is the Infection Control Coordinator at her place of employment, a rural health facility. I interviewed Paula about the uses and challenges of proposal writing.

Q: How do you use proposal writing at work?

A: For program presentations or requested program changes in order to demonstrate merits of these changes.

Q: How has knowledge of this style and understanding of this technique helped you?

A: They provide a systematic, scientifically-based proposal model.

Q: What is your greatest challenge when constructing a proposal or similar document?

A: Often we are in a hurry to complete projects by a deadline. The greatest challenge when constructing a proposal is the time-intensive research.

Q: What tips can you offer someone learning to write these documents?

A: Use an outline. Do not reinvent the wheel!

Q: Describe the context and mission of the particular item you have contributed.

A: My mission was to provide an outline for writing a medium-size proposal on the subject of research/clinical interests in hospital-associated infections. This document is my day's work inside and out.

Q: What was challenging about constructing it, and why?

A: Often it is difficult to cover all areas of a proposal because we might use only components when presenting a particular proposal to an employer or group of people.

Developing her project involved writing several drafts of a proposal. New information and improved understanding of the context of her research led to successive revisions of her document, parts of which are reproduced in Figure 4.7.

Figure 4.7a Proposal by Paula Fields—Introductory Summary

INTRODUCTORY SUMMARY

Handwashing is the single most important procedure for preventing nosocomial infections. Unfortunately, studies repeatedly show that physicians, nurses and others do not always wash their hands before and after all patient contacts. As a result, nosocomial infections remain a cause of morbidity and mortality throughout the world.

According to Bare and Smeltzer (1992) a nosocomial infection is an infection acquired during hospitalization. Nosocomial infections occur in about 5% to 6% of all hospital patients in the United States, accounting for an incidence of 2 million hospital-acquired infections per year. These infections prolong the hospital stay (an average of 13 days for each infection) and represent a direct economic liability of 5 to 10 billion dollars annually. Development of an infection occurs in a cyclical process that depends on the following elements: the infectious agent or pathogen, reservoir, means of transmission or vehicle, portal of entry to host, and susceptible host. Infection develops if this chain stays intact.

(continued on next page)

Figure 4.7a   *(continued)*

The "Chain of Infection" can be described using an analogy. Consider a chain for a vehicle tire. If a link is cut out the chain will no longer stay on the tire. The same principle applies to the chain of infection. If one link is cut an infection cannot be transmitted.

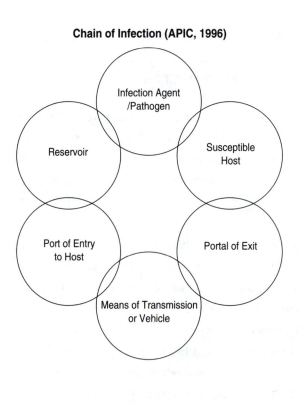

Chain of Infection (APIC, 1996)

Figure 4.7a *(continued)*

Patient health demands interventions to facilitate improved handwashing of physicians, nurses and others before and after patient contact. Improved handwashing would break the chain of infection by eliminating the means of transmission. Improved handwashing would minimize hospital-associated infections, thus reducing related morbidity and mortality. As APIC (1996) states, "Handwashing is the single most important procedure for preventing nosocomial (facility-associated) infections."

Chain of Infection – Broken
(No longer capable of causing sickness)

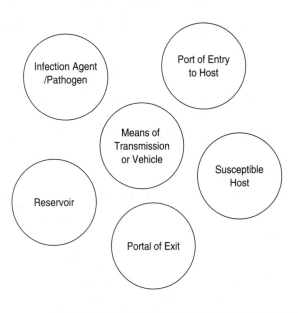

(continued on next page)

Figure 4.7a *(continued)*

> Past research focusing on healthcare individuals' internal motivational factors and attempts to improve compliance have been shown only to temporarily improve handwashing practices, though long-term effects remain minimal. In fact, "No single intervention has been successful in improving and sustaining such infection control practices as universal precautions and handwashing by health care professionals" (Kretzer & Larson, 1998).
>
> Also, Hand-Transmitted Infection notes many interesting points: In health care, nurses and doctors wash only 30% of the required time between patient contacts and procedures. Health care workers, through patient contact, are a leading cause of transmission of nosocomial infection. Nosocomial infection is the most common immediate cause of death in nursing home patients, and the leading cause of patient hospitalization.

Figure 4.7b Proposal by Paula Fields—General Statement of the Challenge

> ### GENERAL STATEMENT OF THE CHALLENGE
>
> If health care workers' training does not effectively or consistently improve handwashing practices, is it possible that educating consumers about the importance of handwashing would more consistently improve handwashing practices? How many healthcare workers could ignore a patient's request for them to wash their hands?
>
> The *purpose* of this study is to examine the effects of healthcare consumers wearing a badge, reading "Did you wash your hands?" and the effect

Figure 4.7b   *(continued)*

of the healthcare consumer's participation in the handwashing practices of healthcare personnel. It is believed that "the education of patients about their role in promoting handwashing compliance among healthcare workers can increase that compliance, as well as provide continuous reinforcement of handwashing principles to healthcare workers, thus promoting long-term handwashing improvements" (McGuckin, 1997).

Although much research has been done in assessing the effectiveness of various interventions regarding the healthcare worker, improvement is only initially noted but long-term effects are minimal. A review of the literature on handwashing has documented the lack of research on the education of patients/residents as an intervention for changing staff handwashing practices. In fact, the only literature available on patient participation in Acute Care (no Long-Term Care data is available) is by Dr. Maryanne McGuckin from the University of Pennsylvania. In 1997 Dr. McGuckin conducted a prospective, controlled, 6-week intervention/control study in four community hospitals in south New Jersey. Patients were educated about the importance of asking their health care workers to wash their hands. The patients' handwashing education model increased soap usage by health care workers an average of 34%. Dr. McGuckin showed for the first time that education of patients regarding their role in monitoring handwashing provides an intervention that indeed reinforces the healthcare worker's washing. Dr. McGuckin's model has shown that patient education may be the key to improve handwashing practices. Dr. McGuckin's model was tested in the Acute Care Facilities for six weeks. The theory is that if her model works in Acute Care then more than likely it will work in a Long Term Care Setting.

Figure 4.7c   Proposal by Paula Fields—Summary Closing the Literature
Review

SUMMARY

The lack of handwashing is a problem that clearly demands the attention of healthcare personnel and healthcare consumers. The studies reviewed
suggest that patients who contract a nosocomial infection have an increased
morbidity and mortality.

Most of the existing research in this area has been monitoring intervention aimed at improving healthcare personnel's handwashing practices.
But, the researchers have focused on short-term effects; in consequence the
research on long-term intervention effects on handwashing practices is minimal. Also, little research has been completed on healthcare consumers' participation in attempts to improve healthcare personnel's handwashing practices, especially in Long Term Care. Clearly, additional research is demanded
regarding the long-term effects of patient participation and increased handwashing compliance in Long Term Care.

Figure 4.7d   Proposal by Paula Fields—Terms Defined

• Bed-days:	The number of days a bed is occupied
• CDC:	Centers for Disease Control and Prevention
• Census days:	Number of days a bed on the Long Term Care Facility was occupied
• Contacts:	Direct interaction between healthcare personnel and healthcare consumers
• Healthcare consumer:	Any person who receives contact/care from healthcare personnel
• Healthcare personnel:	Any individual who has contact with health-care consumers

Figure 4.7d *(continued)*

• Hospital Acquired Infection:	An infection acquired while in a health care organization. The term is interchangeable with Hospital Associated Infection and Nosocomial Infection
• Hospital Associated Infection:	An infection acquired while in a health care organization. Term is interchangeable with Hospital Acquired Infection and Nosocomial Infection¶
• Infectious:	Capable of causing infection or disease (Bare & Smeltzer, 1992)
• Long Term Care:	Referring to a health care organization other than acute care
• Morbidity:	Illness
• Mortality:	Deaths
• Nosocomial Infection:	An infection acquired while in a health care organization. The term is interchangeable with Hospital Acquired and Hospital Associated Infection
• Soap utilization:	Defined in terms of units; each unit represents 800ml of soap

Notes Regarding This Proposal

Introductory summary. This summarizes, defines, and illustrates the problem. The visual material is easily followed and could be employed in presentations at the workplace, using software to make slides for presentation. Notice that this introduction presents disturbing material but does not need theatrics to do so: the facts are dramatic enough without rhetorical embellishment.

General statement of the challenge. Uniting workplace and academic concerns, this section begins with questions to be answered. The purpose of the study, then stated, replies to these questions. This part of the proposal closes with a preview of the literature review that reinforces the problematic aspect of the topic and serves as a transition to that unit.

Summary closing the literature review. This element of the proposal condenses and recapitulates the findings discussed in the literature review, but also closes with a statement reinforcing the need for Paula's research.

Terms defined.   Many people at different levels of technical understanding will read a proposal that unites workplace and academic concerns; hence, terms obvious in meaning to some might be somewhat vague to others. Paula has elected to cover all possibilities in this document.

Retrospective.   This proposal underwent many changes before and after this particular draft as it fulfilled its mission to become a precursor of a workplace plan and of an academic project as well. The streamlined design of the document reflected the need to respond to each constituency. The entire proposal is organized as follows.

◆ *Introductory summary* including a review of the problem
◆ *General statement of the challenge* incorporating the purpose of the study
◆ *Review of literature* surveying what has been and is yet to be done
◆ *Statement of need* focusing on the problem and its relative absence in the literature, a key issue in Paula's project
◆ *Purpose of the study*
◆ *Terms defined*
◆ *Research design* including
    Description of design
    Population
    Method of data collection
    Reliability and validity
    Data analysis
◆ *References*

Advice

Consider these issues in growing such a proposal. Though the segments of proposals and theses can vary depending on academic setting, Figure 4.8 depicts the general process by which raw expectations at the beginning of a project become organized and refined within the proposal stage, and shows how the parts of the proposal correspond to those of the thesis that will emerge from them. Of course, the proposal lacks the content and depth of the thesis and is written in future tense—such and such *will* occur. The thesis describes the research in present and past tense. The introduction and literature review expand; the methodology section discusses what was planned and executed; the expectations of the proposal become the results of the thesis, though their interpretation may need to be altered.

One may compare Figures 4.9 and 4.10 as well note how a five-section proposal can be constructed (Figure 4.9) and how a five-chapter thesis can be developed from that proposal (Figure 4.10). Remember, though, that there are many ways to write such documents and that each depends on the requirements of your field.

Figure 4.8 A Thesis as the Result of a Proposal

Beginnings	**The Proposal**	**The Thesis**
Survey research to create Problem Statement and Research Question	Write the proposal Introduction Literature Review Methods of Research	Write the thesis Introduction Literature Review Methods of Research
	Expectations Interpretation Timetable	Results Interpretation Conclusion

Elements of the project proceed in sequence. Initial survey of research generates a Problem Statement and a Research Question. These items are incorporated into the Proposal. The structure of the Proposal is the model for the structure of the Thesis; the parts correspond, except that Expectations is replaced by Results. Proposals, of course, rely on the future tense, whereas the thesis informs the reader of what has been accomplished.

Figure 4.9   A Detailed Look at a Thesis Proposal

A Detailed Look at a Thesis Proposal

I.   <u>Introduction</u>
(Often the Introduction and Literature Review are combined into one section headed <u>Rationale</u>).

II. <u>Literature Review</u>
- Explain problem, key question inspiring the research, or the concept under scrutiny.
- Review and evaluate previous relevant work.
- Provide a statement of purpose: what will your project do?

III. <u>Methods of Research</u>
- Describe the construction of your project—
  - In a summary,
  - Stating its hypothesis,
  - Enumerating questions your project will answer.

IV. <u>Expectations and Interpretation</u>
- Discuss anticipated data.
- Explain analysis to be performed on data.
- Show the value of the expected information.
- Provide ways of seeing the constraints on project.

V.   <u>Timetable and Contact Information</u>
(This section may or may not be desired by a thesis committee; check with them).
- Provide timetable of completion of all phases of project including the writing.
- Provide contact information for the committee or board.

Figure 4.10   The Thesis Is Organized Like the Proposal

The Thesis Is Organized Like the Proposal

I.   Introduction
(Often the Introduction and Literature Review are combined into one section headed Rationale).

II.   Literature Review
- Explains problem, key question inspiring the research, or the concept under scrutiny.
- Reviews and evaluates previous relevant work.
- Provides a statement of purpose: what did your project do?

III.   Methods of Research
- Describes the construction/methodology of your project—
  - In detail,
  - Stating its hypothesis,
  - Enumerating questions your project answered.

IV.   Results
(Describes sample, data acquired, attempts to falsify the hypothesis, other information generated by the project).

V.   Interpretation and Conclusion
- Discusses results of data.
- Explains their meaning and place in context of other research examined.
- Explains the constraints of the project.
- Proposes future studies overcoming such constraints.
- Explains value of results from practical and theoretical perspectives.

For Further Study

1. How do long proposals reflect the style and strategy of shorter prospectuses, particularly in the merging of writing styles across the disciplines?

2. If you were to produce an electronic presentation to accompany a written formal proposal, what choices in selecting content for display would you make?

3. Would there be the need to modify the headings in the written document in order to construct appropriate slide titles?

4. If you gave an oral presentation of your proposal using computer-generated material, how would it integrate the content of the written document and that of the computer-generated presentation?

Review of Patterns

Experimental Project

Introduction

Statement of purpose or problem

Rationale

Hypothesis

Stipulations

Definitions

Literature review

Discussion of method

      Equipment

      Design

      Ingredients

      Evaluation

      Analysis

Timetable

Appendix

References

Qualitative Proposal

Introduction

Significance or rationale

Role and scope of project

Definitions

Literature review

Methodology

Expectations

Timetable

Appendix

References

Simplified Long Proposal

**Prospectus for an
Academic Project**

Abstract (if required)

Introduction

Problem

Hypothesis

Project method

Expectations

Timetable

Appendix

References

Or:

Abstract or summary

Introduction

Rationale

Methodology

Staffing

Budget

References

Appendix

**Business or Institutional
Proposal**

Executive summary

Introduction

Statement of challenge

Statement of agenda

Means to address challenge

Expected results

Timetable

Supporting material

References

An Article-Length Proposal

Title page

Abstract or executive summary

Table of contents (and list of illustrations, if necessary)

Introduction

Statement of significance

Role and scope

Definitions

Literature review

Theoretical basis

Hypothesis

Methodology

Expectations and implications

References (may come last)

Appendix

A Short Thesis Proposal

Introductory summary

General statement of the challenge

Review of literature

Statement of need

Purpose of the study

Terms defined

Research design

      Description of design

      Population studied

      Method of data collection

      Reliability and validity

      Data analysis

References

5

Grant Proposals— Some Suggestions

This chapter surveys some considerations in writing grant proposals. No justice can be done to the nearly infinite breadth and variety of such grant applications, but this text can present analysis of common examples and the philosophy of constructing this form of communication—a form both businesslike and scholarly.

Questions

◆ What are successful strategies of grant proposals?
◆ What stylistic aspects must the writer of such documents understand?
◆ What needs dictate the presentation of content?
◆ ·What do these proposals borrow from other documents we have seen?

RFP, Application, Preproposal.   The prospectus for grant funding usually responds to a call for proposals or a request for proposals (an RFP) posted in professional literature, on the Internet, or in mailed documents (Figure 5.1). Rarely will a grant proposal be unsolicited; though there are exceptions, the shape and content of most grant proposals derive from guidelines presented by grantors. Many grant-making organizations are conservative in matters of format and style, and expect that you will closely pattern your argumentation and development upon a structure determined by the requirements in the RFP or application packet.

Normally the RFP itself will present the criteria for constructing your prospectus, or else your preliminary response to it will result in application materials being sent to you. These will contain instructions for creating your document (see Figure 5.2). Because the RFP or the application will specify the scope of the grant funding and the nature of the organization, you will usually know or be able to research the potential grantor to determine the needs, the readership of your proposal, and the examples of projects that have been approved. Sometimes the writer must respond to the RFP with a *preliminary*

Figure 5.1   Contents of an RFP

Contents of an RFP

A Request for Proposal may specify that your proposal or preproposal include the following elements:

- <u>Cover sheet</u> that is preprinted by the requesting organization
- <u>Descriptive narrative</u> limited in size to a certain number of pages and explaining the project's
  - Scope
  - Significance, and
  - Procedure or methods
- A <u>summary explanation</u> of the importance of the project to the organization funding the award
- A <u>summary explanation</u> of the importance to the sponsoring governmental entities
- A <u>budget statement</u> using allocation forms provided by or specified by the grantor
- <u>Vitae or resumes</u> of all principals involved

Figure 5.2   Contents of an Application

Contents of an Application

The application may contain elaborations of elements found in an RFP. For example, you may have to provide:

- Detailed information on a special cover sheet
- An abstract explaining the significance of the project
- Budget summary on allocation forms
- Project description
- Biographies of principals, and their resumes
- Certifications of eligibility and compliance with regulations

---

proposal that will be reviewed in an initial round. If reviewers respond positively to that preproposal, the writer will then be invited to submit a detailed proposal.

In other cases review of a *single* submitted proposal proceeds in stages, the first being a screening and the second a detailed reading—if the document passes the screening review (see Figure 5.3). Last year's set of standards might differ significantly from the one the organization is using this year, so you must

Figure 5.3   Proposal Creation and Review Process

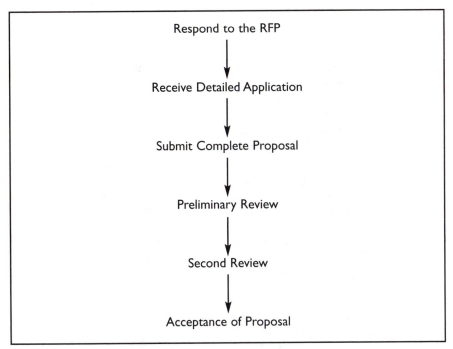

Respond to the RFP

Receive Detailed Application

Submit Complete Proposal

Preliminary Review

Second Review

Acceptance of Proposal

be sure that your information about the granting entity is current. Naturally, proposal requirements vary with the grantor and the field, so you should never assume that one version of the prospectus answers the needs of all grantors. For example, if you apply to two organizations for the same project needs, you will have to phrase and build your two proposals differently—even if one organization is a subset of the other. That sounds rather obvious, but as I surveyed grantors, I noted this as one of their concerns.

Applications

Whether they are short or long, grant proposals require that you

◆ Fine-tune your prospectus to reflect the vocabulary, stylistic expectation, and value system of the reading committee. You will need to use the grantor's terminology in discussing your project, the grantor's schematic for depicting budgetary outlay, and the grantor's preferences in presentation. You must make the case that your project aligns with the grantor's overarching goals.

◆ Stay within the length and adhere to the format specified in the application. If the instructions describe a five-page, double-spaced depiction of the project, then that is the length the narrative should be. If the application contains required worksheets, then use these in the proposal. Most such items can be scanned so that you can type in the spaces required—or you can paste up your typed material neatly, and then photocopy the whole so that a clean-looking page results.

◆ Comply with the strictures of technical writing. Sloppy third-generation copies, hard-to-read text, and unorganized sections will alienate the grantor. Application materials usually specify headings to be used; use them. Perfect spelling and punctuation is vital.

◆ Detail the proposed project so that the review committee does not raise awkward questions about omitted material. If your budget is not clear, or if your narrative does not say how the activities of your project are consonant with its goals and with the grantor's objectives, your proposal will not be considered.

◆ Plan realistically so that you can adhere to the timetable explained in the proposal. This will prevent the considerable ill will that will occur if you receive the grant and then have trouble following through.

◆ Contact the grantor's staff if you have any questions about any part of the application process. These personnel are usually helpful, and normally grateful that you have consulted them.

In grant applications, then, the text must inform the reviewers and persuade them to appropriate money or other resources. Grantors usually have limited resources and many applicants who desire funding. So scrupulous attention to instruction and detail is mandatory.

Example 1. A Proposal to a Humanities Foundation Requesting Funding for a Research Project

What It Looks Like.   Figure 5.4 depicts a grant proposal that was not written to stand by itself—rather, the granting agency required a large packet of supporting material to accompany this proposal: worksheets and an extended resume or vita. The text as shown appears after a cover sheet containing contact information and an abstract of the proposal. Following the text, I attached a vita with references.

Figure 5.4   Text of Short Grant Proposal

PROPOSAL

TO: The West Virginia Humanities Council
FROM: Dr. Brian R. Holloway, Associate Professor of English
              The College of West Virginia
DATE: December 30, 1995
RE: Proposal for Fellowship in the Humanities

Summary

I request a Fellowship award of $2000 to assist research that would produce original scholarship and enhance the courses I teach. My request is consonant with the current *NEH Initiative on American Pluralism and Identity,* as my project overlaps literature and the study of interacting cultures. I propose to research the work of John Neihardt and Nicholas Black Elk in creating *Black Elk Speaks.* This book has exerted worldwide, multicultural influence while dramatically affecting Native American concerns and studies. Yet though it is frequently taught—and quoted—its context and origin remain unclear, the subject of debate.

Proposal: *Intermingled Cultures—*
John Neihardt and Nicholas Black Elk

I shall use the 1996 stipend to travel to the University of Missouri–Columbia to work with the John Neihardt collections at the Western Historical Manuscript division of the Elmer Ellis Library for four weeks during the summer of 1996. I will research primary materials and artifacts produced by Neihardt's collaboration with Black Elk—a project which resulted in *Black Elk Speaks.* By examining the firsthand evidence and its context, I intend to determine the extent to which the book was a collaboration of two cultures or a compromise between them.

I have already begun this project by contacting the librarian in charge of the Western Historical Manuscript Collection at the Elmer Ellis Library. This collection houses Neihardt's original manuscripts, field notes, supporting materials, and letters concerning *Black Elk Speaks.* Also on that campus is the Neihardt Collection in the English Department's headquarters in Tate Hall; it consists of the personal library of John Neihardt, which contextualizes Neihardt's work. These collections will be available throughout the summer of 1996, when I plan to see them. Permissions will have to be granted to use or quote certain restricted materials. Though limited inspection of microfilms of certain materials through interlibrary loan is possible, it is important to examine the primary evidence, not transcriptions, for insight into the construction of Neihardt's book.

(continued on next page)

Figure 5.4 *(continued)*

2

My Objectives in Studying the Firsthand Materials Are:

To provide a scholarly basis for interpreting *Black Elk Speaks* in the college classroom, so that it may be used successfully in either a World Literature, Native American Literature, or stand-alone course;

To generate articles, culminating in a book, which will so inform others in the academic community (currently there are several contradictory books which obscure, rather than assist in, interpreting this text);

To integrate the scholarship and the text itself into English 210 (World Literature II) on our campus, a course which I have developed;

To create a separate course using *Black Elk Speaks* as the primary text. I envision a course immersing students in cultural studies, *Black Elk Speaks,* and analysis of the current influence of this work. (The "messianic" aspects of the Ghost Dance itself have a lot to tell us about long-unstable social conditions, just as Black Elk's philosophy exerts strong cross-cultural influence);

To publish the resulting course on the Internet so that other colleges will benefit from this project.

Consistency

This project derives from long association with secondary material. My involvement began in 1974 when, for high-school students, I taught Neihardt's *When the Tree Flowered* and *Black Elk Speaks.* I have used these books in advanced rhetoric courses at the University of Illinois, and have presented them in modern literature and lifelong learning classes at Parkland College. Here at The College of West Virginia, I incorporate *Black Elk Speaks* into the new offering of English 210 (World Literature and Cultures II) which I recently designed. In addition, a paper on "World Literature and the Canon" which I delivered at the fall 1994 WVACET conference discussed the book Neihardt and Black Elk wrote; this book emerged as a topic in my presentation at the Lilly Conference on Excellence in College Teaching November 17, at Miami University, Oxford, Ohio: "Enhancing the Options—Cross-Cultural Studies in the Literature Classroom." That discussion has been submitted to the *Journal on Excellence in College Teaching.*

I now wish to use the primary materials to further my study of Neihardt and Black Elk, and to benefit The College of West Virginia. I have enclosed a brief bibliography, a budget estimate, and an abbreviated resume for your consideration. I welcome your support.

Figure 5.4   *(continued)*

3

Bibliography

Black Elk, Nicholas, *The Sacred Pipe: Black Elk's Account of the Seven Rites of the Oglala Sioux*. Ed. Joseph Brown. Norman: U of Oklahoma P, 1953.

—. *The Sixth Grandfather*. Ed. Raymond DeMallie. Lincoln: U of Nebraska P, 1984.

Castro, Michael. "John G. Neihardt." *Interpreting the Indian: Twentieth Century Poets and the Native American*. Albuquerque: U of New Mexico P, 1983.

Deloria, Jr., Vine, ed. *A Sender of Words: Essays in Memorial of John G. Neihardt.* Salt Lake City: Howe, 1984.

Erdoes, Richard. "My Travels with Medicine Man John Lame Deer." *Smithsonian* 1973: 34.

—. *Crying for a Dream*. Santa Fe: Bear, 1990.

McCluskey, Sally. "Black Elk Speaks: And So Does John Neihardt." *Western American Literature* 6 (1972): 231-242.

Rice, Julian. *Lakota Storytelling: Black Elk, Ella Deloria, and Frank Fools Crow*. New York: Peter Lang, 1989.

—. *Black Elk's Story*. Albuquerque: U of New Mexico P, 1991.

Sayre, Robert. "Vision and Experience in *Black Elk Speaks*." *College English* 32 (1971): 509-535.

Budget

Transportation	(2 x 660 miles @ .26/mile)	343.20
Meals	($10/diem)	280.00
Lodging	($145/week)	580.00
Permissions	(Estimated 16 @ $50)	800.00
Book purchases to date		170.58
Research/Writing—min. 40 hrs./week @ $25/hr.		4000.00
Total expenses, estimated		6173.78
Offset, Fellowship		<2000.00>
Contribution in-kind		4173.78

How It Works. Note that the text of this prospectus uses a modified memo style but deviates slightly from the familiar proposal format, per application guidelines. Indented first sentences of paragraphs and centered headings provide a balanced, symmetrical look. Bold headings signal the beginning of key sections, and the contrast between normal and italicized type in the second heading emphasizes the concept of the project. The support section summarizes the proposed project, states objectives in an outline form, and explains qualifications (the "Consistency" material). A closing paragraph follows, requesting support for the endeavor. Because the bibliography and budget were required attachments to the end of the prospectus, I wanted to insert the closing request before readers could become distracted by those items. I had the option to do this since the application materials contained no restriction concerning such an organizational choice. I decided that I would use a similar pattern to apply for support for the second phase of this research.

Example 2. A Follow-Up Proposal for the Next Round of Research

What It Looks Like. Earlier research led to more research and to another grant request. By this time, the format of presentation had changed slightly, though the grantor still provided basic forms that needed to be used. There was the cover sheet with contact information, project title, and abstract. There was the requirement to include a resume at the end. In Figure 5.5, I have deleted these items to focus on the text itself, what applications often call the "narrative."

Figure 5.5 The Second Grant Proposal in the Series

PROPOSAL

TO:	The West Virginia Humanities Council
FROM:	Dr. Brian R. Holloway
DATE:	November 23, 1998
RE:	Proposal for Fellowship in the Humanities

Summary

I request a Fellowship award of $2500 to assist research that would produce original scholarship and enhance the courses I teach. I propose to research the work of John Neihardt, Nicholas Black Elk, Eagle Elk, and Andrew Knife in creating *When the Tree Flowered*. This book—a compendium of Lakota stories, history, and cultural knowledge woven together as a narrative intended to educate the dominant culture—has exerted worldwide, multicultural influence

---

Figure 5.5   *(continued)*

while affecting Native American studies; the text has had many imitators as well. Unlike *Black Elk Speaks*, *When the Tree Flowered* has not been directly embroiled in partisan controversy driven by religious and political agendas. Yet it has been produced in much the same way as its predecessor, and the process of its creation has also been seriously misunderstood—both by critics and by those copying the book's format and intent. A scholarly understanding of this process will illuminate not just the context of *When the Tree Flowered*, but also other cross-cultural works of this kind.

Some of the basic questions I hope to research are these:

—What, if anything, had changed in the Federal Government's attitude toward Neihardt and his meetings with Lakota elders? Why was there apparent resistance in 1931, when Neihardt was the "outsider" looking in? Was there a subtle institutional resistance in 1944 once Neihardt became a government worker under Collier? How did government attitudes toward Neihardt affect his work in 1931 with *Black Elk Speaks* and 1944, with *When the Tree Flowered*? I am convinced documents exist which will illuminate these issues; Hilda Neihardt probably has anecdotal material which will help, as well.

—Is there—as some critics have suggested—a shift in Neihardt's depiction of the Lakota, or is *When the Tree Flowered* of a piece with its famous predecessor? Do both works, in fact, serve as a synecdochic portrayal of the life of a people as imaged through one main individual? Studying the 1944 transcriptions and subsequent draft manuscript will assist in determining this.

—How effective was the process by which notes were taken which became the 1944 manuscript? To provide the material for *When the Tree Flowered*, a format was used similar to that by which Black Elk and others were interviewed in 1931—with the exception that in 1944, Hilda Neihardt typed the notes in the field on a portable typewriter as Ben Black Elk translated them (in 1931, Enid Neihardt, the oldest daughter, took stenographic notes and then transcribed them). What—if any—difference did the two systems make? Manuscripts, notes, and interviews with Hilda Neihardt will clarify these points.

—The young Neihardt was very familiar with the social customs and language of the Omaha (who spoke a Siouan language, though originally at war with the Sioux). How did this affect the mature Neihardt's understanding of Lakota ways and communication? Interviews with Hilda Neihardt and archival material will help to illuminate this.

—Does *When the Tree Flowered* contain traces of the Marxist dialectic, mystical neoPlatonism, or puritanical religious fervor some have accused *Black Elk Speaks* of harboring?

(continued on next page)

Figure 5.5 *(continued)*

Proposal: *The Harvest from the Flowering Tree—* John Neihardt's 1944 Interviews

I shall use the 1999 stipend to:

1) Travel to the Spring Neihardt Conference, at which issues of orality and literature will be paramount (N. Scott Momaday will speak, and students at the Red Cloud School at Pine Ridge will discuss casting oral tradition into written form). At the conference, I shall interview Hilda Neihardt and others, including (I hope) N. Scott Momaday, who has written about oral traditions and Neihardt. I shall view and listen to archival tapes at the Neihardt Center. Since I will be driving, I can stop at the University of Missouri-Columbia, to work with the John Neihardt collections at the Western Historical Manuscript division of the Elmer Ellis Library. There, I will research primary materials and artifacts produced by Neihardt's collaboration with Eagle Elk, Andrew Knife, and Black Elk.

2) In early summer, I will return to the Neihardt Center by way of the University of Missouri-Columbia for an extended two-week cycle of research involving archival material and interviews. Because of my previous work with *Black Elk Speaks* (partially funded in 1996 by a WVHC grant), I have already been in contact with the Western Historical Manuscript Collection at the Elmer Ellis Library. This collection houses Neihardt's original manuscripts, field notes, supporting materials, and letters. Also on that campus is the personal library of John Neihardt, which contextualizes Neihardt's work. Though limited inspection of microfilms of certain materials through interlibrary loan is possible, it is vital to examine the primary evidence, not surrogates, for insight into the construction of Neihardt's books. As I discovered during my previous research, the original manuscripts contain a variety of evidence not apparent from examining film copies; moreover, much of the original writing and emendation is written in pencil and does not reproduce well, or at all. Raymond J. DeMallie's published transcriptions, though helpful, have been edited, regularized, conflated, and adjusted on the page for a general readership and diverge in important ways from the original texts.

My objectives in studying these first-hand materials are:

To provide a scholarly basis for interpreting *When the Tree Flowered* in the college classroom, so that it may be used successfully in either a World Literature, Native American Literature, or stand-alone course;

To integrate the scholarship and the text itself into World Literature II, American Literature II, Interdisciplinary Studies Capstone Seminar, or a pending "Special Topics" course on our campus;

Figure 5.5   *(continued)*

To generate articles, culminating in a book, which will assist others in the academic community (currently there are several contradictory books which obscure rather than assist in interpreting this text);

To share course syllabi using *When the Tree Flowered* through the annual Western Literature Association *Syllabus Exchange*, so that other colleges and teachers will benefit from this project. The *Exchange* is a published anthology of syllabi and commentary.

Consistency

This project derives from long association with secondary material. My involvement began in 1974 when, for high-school students, I taught Neihardt's *When the Tree Flowered* and *Black Elk Speaks*. I have since used these books in many college courses over the years: courses including Advanced Rhetoric, Introduction to Modern Literature, World Literature, American Literature, and a Senior Capstone Seminar. In April, 1996, a WVHC grant enabled me to return to the Ellis Library and study archival materials related to *Black Elk Speaks*; I also traveled to Nebraska to interview Hilda Neihardt and to research at the Neihardt Center, staying there as a guest of the family. I have returned in 1997 and in 1998, and was one of three principal speakers at the Spring Neihardt Conference this past April. I have written two articles about my earlier research, and a book which is currently under press review. My most recent presentation using that material, at the October 1998 WVACET conference at North Bend State Park, discussed Neihardt's imitators and their relationship to the tradition founded by his work with his Lakota friends.

I now wish to use the primary materials to further my study of Neihardt, Black Elk, Eagle Elk, and Andrew Knife. I have enclosed a brief bibliography, a budget estimate, and an abbreviated resume for your consideration. I welcome your support.

Bibliography

Black Elk, Nicholas. *The Sacred Pipe: Black Elk's Account of the Seven Rites of the Oglala Sioux*. Ed. Joseph Brown. Norman: U of Oklahoma P, 1953.
—. *The Sixth Grandfather*. Ed. Raymond DeMallie. Lincoln: U of Nebraska P, 1984. Brumble, H. David. *American Indian Autobiography*. Berkeley: U of California P, 1990.
Castro, Michael. "John G. Neihardt." *Interpreting the Indian: Twentieth Century Poets and the Native American*. Albuquerque: U of New Mexico P, 1983.
Erdoes, Richard. *Crying for a Dream*. Santa Fe: Bear, 1990.

(continued on next page)

Figure 5.5 *(continued)*

Erdoes, Richard, and John Fire Lame Deer. *Lame Deer, Seeker of Visions.* New York: Simon and Schuster, 1972.

Irwin, Lee. *The Dream Seekers: Native American Visionary Traditions of the Great Plains.* Norman: U of Oklahoma P, 1994.

McCluskey, Sally. "Black Elk Speaks: And So Does John Neihardt." *Western American Literature* 6 (1972): 231-242.

Momaday, N. Scott. "To Save a Great Vision." in Deloria, Jr., Vine, ed. *A Sender of Words: Essays in Memorial of John G. Neihardt.* Salt Lake City: Howe, 1984. 30-38.

Rice, Julian. *Lakota Storytelling: Black Elk, Ella Deloria, and Frank Fools Crow.* New York: Peter Lang, 1989.

Budget (Both Trips Together)

Transportation	(2 x 2500 miles @ .27/mile)	1350.00
Meals	($14/diem, 3 weeks)	294.00
Lodging	(8 motel nights @ $75)	600.00
	(10 days at UM-C dorm)	230.00
Parking at UM-C		010.00
Copies, slides, permissions, etc.		200.00
Office supplies		050.00

===
===

Total expenses of trips, estimated	2734.00
Offset, Fellowship	<2500.00>
Difference	234.00

How It Works. Again, the application materials dictated length and overall format. I tried to make the sections of the new proposal correspond to those of the original proposal, though, to emphasize that the two research projects were actually part of the same extended endeavor. Both proposals contained a "project narrative" explaining the work to be done and a "budget narrative" explaining how the work would be paid for. (For longer proposals, a name such as "project narrative" may be used loosely in the application to refer to a description of the rationale, problem, literature, methodology, and references). Figure 5.6 depicts the type of project and budget narrative often occurring in shorter arts or humanities proposals.

Figure 5.6   Project and Budget Narrative

PROJECT NARRATIVE

Summary

We request a mini-grant of $1500 twice a year to assist the College of Fredonia Creative Writing Project. This project would cultivate an interacting community of creative writers in southern Fredonia, produce two creative writing anthologies per year, sponsor the reading of original works by local writers in public venues, post the work of such writers on the Internet, and support visits by local writers to regional schools or other sites. The project would continue and enhance the present work of the College of Fredonia Creative Writing Group which fosters publication of and readings by local writers.

Current Project

At present, the College of Fredonia Creative Writing Group sponsors creative writing in our region. Since its inception in 1995, the Group has met regularly during the academic year, reviewing, critiquing, and encouraging work by members and contributors from southern Fredonia. The Group holds readings by local writers at Chic-O's, a nearby coffeehouse, and at the Groucho Marx Memorial Theater. The Group edits, plans, and produces the annual magazine *Duck Consomme* which is distributed free of charge on campus, at coffeehouses, and at libraries, stores, and other locations throughout southern Fredonia. The release of the 2002 volume, notable for its production values and writing, was an important event for the humanities in our community.

This volume featured writers from Harpoville, Zeppotown, and other nearby areas. Both volumes were produced under the direction of Dr. Arthur Schmendrick, Associate Professor of English at the College of Fredonia, who donated his time. The College provided color and black-and-white copying, computers, typing and meeting facilities, and paper, while a small grant by Dr. Schmendrick paid for the binding of presentation copies which contributors received. All 1500 units were assembled and distributed by the Creative Writing Group, and mailed without charge to interested parties. At the same time Fredonia posted the magazine at its Internet site. Without such extensive involvement by the campus, there would have been no publication. Now, following the success of that project, the Creative Writing Group wishes to do more.

(continued on next page)

Figure 5.6 *(continued)*

Proposal: *The Fredonia Creative Writing Project*

We propose that the Creative Writing Group use two $1500 stipends per year to enhance what has been established. Already, the next volume of *Duck Consomme* is in production, and the group will fabricate covers soon which again will feature landscape photography by a local resident. Local readings are planned. But there is much more that the Group would like to do. Its objectives are listed below.

Schedule and Objectives for the Creative Writing Project

To outsource the printing of *Duck Consomme*, so that production values may be achieved without burdening an all-volunteer staff with the mechanical phases of the project. Relieved of these chores, the Group can focus on writing and editing.

To generate two such volumes per year beginning in 2002—one retaining the present eclectic mix to be published in the Spring and one having a thematic emphasis to be published in the Fall. For example, a collection of writing for and by children might come out before the holiday season, or a volume of local ghost stories might be published for Halloween.

To integrate the Group fully into the community by sponsoring appearances of local writers in area schools, at festivals, and at other functions. Two to three readings in this Fall and the Spring would work best.

To continue publishing *Duck Consomme* on the Internet, so that our writers will benefit from global exposure. By the end of 2003, all volumes should be available online.

Biography of Arnold R. Schmendrick, Director of Project

Born in Flushing, New York in 1949, Arnold R. Schmendrick received his B.A., M.A., and secondary teaching certification from Mallard University. He earned his Ph.D. in English from the University of Dogbane in 1979, continuing to teach there afterwards and being cited by students and administration as an outstanding educator of undergraduates. He developed and taught traditional and alternative courses at Panatela College while working on its community writing project and publication, *Zither*, and is now Associate

Figure 5.6 *(continued)*

Professor at The College of Fredonia where he teaches writing and literature and advises the school's literary magazine. In 2000, that college's Student Government Association and its Alumni Association both voted him Teacher of the Year.

Professor Schmendrick has published and presented on modernism in literature, Shakespeare's plays, and Renaissance iconography. Dr. Schmendrick has received numerous honoraria and grants to support extended research. In addition, Schmendrick has long been associated with creative writing—he has participated in poetry readings from the 1970s to the present, teaches the subject at the college level, and was founder of the Creative Writing Group in 1996. His accomplishments also include successful short-story publications in 1977, 1999, and 2000 as well as first prize in the University Series of Aging Poets Contest in 1971.

At the College of Fredonia, Dr. Schmendrick incorporates creative writing into his sections of World and Children's Literature and currently teaches two classes that are each working on collaborative projects of original fiction.

Exhibits

Please see the following items enclosed: volumes of *Duck Consomme*; home site of *Duck Consomme* on the Internet; vita of Arnold Schmendrick.

Plans for Evaluation of Project

Public presentations such as readings can be evaluated using the Humanities Foundation standard form; *Duck Consomme* itself can be evaluated by soliciting reader feedback in the issue and recording the results.

A Need to Reinforce the Humanities

Our creative writing project is by definition an expression of the humanities operating in the lives of ordinary people, and uniting them despite very different backgrounds and career needs. Our humanities emphasis enhances vocational necessities and aspirations. For instance, the current *Duck Consomme* staff includes several business majors—our photography director (just retired and now returning to school), our art coordinator (who wants to open a day-care facility), and the secretary of the Creative Writing Group itself. Please help our Creative Writing Project assist these motivated people, and others like them, to achieve their goals.

(continued on next page)

Figure 5.6   *(continued)*

BUDGET NARRATIVE

Printing costs   (2 x 1500 issues @ 1.00 each)   3000.00

In-Kind Support

Clerical (typing, mailing)   (20 days @ $50/diem)   1000.00
Space (use of facility)   ($10.00/week x 30 weeks)   0300.00
Postage   (Estimate .32 x 200 issues)   0064.00
Advisor   ($25/hr x 2 hrs/week x 30 weeks)   1500.00
Internet expertise   ($50/hr x 4 hrs for 2 vols.)   0200.00

Total expenses, estimated   6064.00
Offset, 2 minigrants/year   <3000.00>
Contribution in-kind*   3064.00

- -

*Does not include travel expenses incurred by participants, as these are
impossible to estimate at this time.

Example 3. A Final Report Discussing Accomplishments of and Questions Raised by the Grant Sequence

What It Looks Like.   After the second round of research, I provided a final
report explaining what had been achieved and what had yet to be accomplished (see Figure 5.7).

Figure 5.7 Final Report for Second Phase of Research

Grant #2422 Final Report

Brian Holloway

Overview of Project

During the grant period I made three extended research trips and continued my research by correspondence. I visited the Western Historical Manuscript Collection, the John Neihardt personal library, the Museum of Anthropology, and the general holdings of the Ellis Library (all at the University of Missouri-Columbia), and the Neihardt Center in Bancroft, Nebraska. This work resulted in the completion and publishing of a journal article, the presentation of two papers, and the design of a syllabus for an upper-division course. These documents are attached to this report. My research on John Neihardt's writings received funding from a West Virginia Humanities Council grant in 1996, continued in a greatly-expanded way once the grant had expired (largely at my own expense, but including a small speaking honorarium from the Neihardt Foundation) and then received funding again in 1999 from the West Virginia Humanities Council. The scholarship itself has evolved throughout this time period, and my research on *Black Elk Speaks* and *When the Tree Flowered* cannot be divided into "periods" of research on separate books at all—rather, I have looked at both topics together, examining issues common to both. I have combined archival work, research with primary and secondary sources, and interviews to create an understanding of the context within which these two books were created. Of course, I have used all the funds allocated for project #2422 in doing so.

Summary of Findings

Many of the issues of context are the subjects of the attached documents. For example, "*Black Elk Speaks* and Some Discontents" reviews the tendency of some to classify Neihardt's Indian writings in categories for which they were not intended, and then to criticize those writings for not living up to such artificially-imposed parameters. Neihardt, faced with the task of presenting the eloquence of Black Elk and his friends in terms that the dominant culture would also recognize as eloquent, was not interested in writing ethnography devoid of poetics. Nor did Neihardt bring to his task a set of simplistic preconceptions sometimes fastened on him by critics.

(continued on next page)

Figure 5.7   *(continued)*

Neihardt's poetics, in fact, are the subject of my paper "Wordsworth, John Neihardt, and Pleasure's Fortunes" which explores the Neihardt-modernist rift and posits a connection between Neihardt and Wordsworth, rather than a link with writers of Puritan heritage or temperament as alleged by some critics. Neihardt's concepts of poetry assisted him in writing what we might call his free-verse epics, *Black Elk Speaks* and *When the Tree Flowered*, and ensured their appeal to the dominant culture. It is interesting to note that the modernist poetry much prized by some academicians but seldom gaining a large general readership held little allure for Neihardt, and that a large rift developed between Neihardt and Harriet Monroe.

There is also the rather conspicuous textual issue—an appropriation by successor writers of the very language created by Neihardt to tell the stories of *Black Elk Speaks* and *When the Tree Flowered*. Neihardt's work with Black Elk, Eagle Elk, Andrew Knife and others, and his sensitivity to Siouan languages engendered by familiarity with the Omaha in his youth, created a style much imitated by others who followed in his path—whether or not those authors acknowledged their debts. Indeed, one cannot watch a film about the Plains wars nor read current paperback books about that time and that conflict without being aware of phrases, expressions, and modes of telling that owe their popularity or even existence to Neihardt's collaborations. My paper "Illicit Text?" discusses this continuation of the style and manner made notable by Neihardt in *Black Elk Speaks* and *When the Tree Flowered*. In a sense, Neihardt's work with his Lakota friends helped create an enduring American prose-epic language.

There is the additional question of how the material of these prose epics should be taught. I have incorporated these books into sections of American Literature, World Literature, and Senior Capstone courses. I have tried to ensure that students are exposed to the context informing these books and the cultural issues surrounding them by providing bibliographical information within syllabi and by assigning students group project work requiring the use of the Internet's ever-growing resources on Plains culture and John Neihardt. My recent syllabus is attached.

In November 1998, I developed as a focus of research the short-range goals outlined in my proposal for Grant #2422, phrased in questions that would certainly undergo modification as I became familiar with more primary material. This portion of the report responds to those concerns listed in my original proposal and not previously addressed in this document.

First, I found no written evidence of any inhibiting governmental attitude impinging on Neihardt during the time of the 1944 interviews that produced *When the Tree Flowered*. This contrasts with the federal bureaucracy's odd attitude to Neihardt in 1931, which was complimentary on the surface and apparently secretly wary, as evidenced by several pieces of correspondence in the Neihardt Collection. Collier's administration, which superseded the old regime, probably affected the way bureaucrats perceived Neihardt. Of interest, however, and remarked on by the Neihardt family, is the fact that Neihardt's publisher seems to have expressed continued reluctance regarding the author's proposals for books about the Lakota, and a hoped-for project about the Paiute

Figure 5.7 *(continued)*

Wovoka, originator of the " Ghost Dance." Hilda Neihardt has told me that her mother insisted to John Neihardt that his publisher was stifling him.

Studying the draft of and notes for *When the Tree Flowered* reveals their similarities—the same interpreter (and Hilda Neihardt states that Ben Black Elk spoke good English), a process of discussion among the tellers, generating a collaborative narrative, a Neihardt taking notes (in this case, Hilda, using a portable typewriter). A multiple account included in the transcriptions for *When the Tree Flowered* becomes a unified telling in this later work, which is a "nonfiction novel." The process by which notes were taken and the book created seems to have been very effective. Other initial questions have been discussed previously in this report.

Evaluation and Conclusion

The material I have encountered will undoubtedly take years to assimilate. I have no illusions that I can write rapidly and get published quickly a large critical exposition on Neihardt. In part, that is because the material itself requires much consideration and my academic job demands much time. In part, that is because the critical view dominating some presses and journals may resist a balanced reappraisal of Neihardt. For example, I sent an editor a meticulously-written and carefully-documented prospectus. It was returned with a rejection note containing grammatical solecisms, accompanied by a review that could not fault scholarship but that criticized the submission for employing the semicolon and for using adverbs. In addition, the reviewer wanted a product with a different agenda. My correspondence with several other scholars of Neihardt reveals a similar pattern in which reviewers cannot dispute facts but instead insist on a different manuscript fulfilling the reviewers' preconceptions. These agendas may reflect issues that I discuss above when surveying "*Black Elk Speaks* and Some Discontents," or they may simply result from what critic Frank Lentricchia and commentator Deborah Tannen have noticed in academic culture. To Lentricchia ("Last Will and Testament of an Ex-Literary Critic," *Lingua Franca*, September/October 1996, 59- 67), critics' agendas often derive from "theory" precedent to an examination of the work: in fact, it is possible to pronounce a great many things from a theoretical perspective without ever having read the work itself. Indeed, the primary artifact's realities may prove to be impediments to such critical pronouncements. Tannen, in a recent article in *The Chronicle of Higher Education* ("Agonism in the Academy: Surviving Higher Learning's Argument Culture," 3-31-00, B7-8), notices the same phenomenon, and relates it to what she perceives as a ritualized contention embedded in the academic construct—a stultifying force inhibiting scholarly approach. Having been an editor myself, I incline to the view that academic outlets for publication succumb mostly to inertia, rather than to ideology or querulousness, and I believe it will take years for such inertia to dissipate. I therefore have plenty of time to refine my understanding of the primary materials through study and teaching, and to achieve better results when synthesizing my findings in any eventual large publishable document. Overall, this project accomplished that which it set out to do, and for the continued support of the West Virginia Humanities Council I will be forever grateful.

How It Works.   This document, required by the grantor, brought closure to the second phase of the project while indicating that much more work remained to be done. Of interest is that my final report for phase one had been more optimistic. By the end of the next round of research, however, I realized that I had probed the perimeter of what could easily be a life's work. I wanted to be realistic in this document, and still leave open the prospect that I might apply for future assistance with this project.

Example 4. Some Components of a Major Proposal for a Scientific Grant

What It Looks Like.   Figures 5.8, 5.9, 5.10, 5.11, and 5.12 contain sections of a draft of a formal grant application by Douglas M. Burns, a microbiologist. This large, book-length document summarizes existing research and proposes a new project. The complete report's many parts derive from technical format, though driven by the protocol of work in a specialized field.

Figure 5.8   Abstract: Part of Proposal by Douglas M. Burns

Reports from our laboratory and from several other active groups strongly suggest that the neuropeptide kalrectin gene-related peptide (KGP) stimulates osteoblasts. Additional reports demonstrate osteogenic actions and in vivo skeletal effects of KGP. KGP is abundant in skeletal areas and bone marrow stromal regions, where it is thought to act as a neuroeffector in the maturation and physiologic regulation of osteoblastic cells. We, and several other well-respected international laboratories, initially demonstrated direct stimulation of osteoblast proliferation and direct stimulation of cAMP production in osteoblasts at 500-fold lower levels than those necessary to produce effects in osteoclasts. Bernard and Shih reported that exogenous KGP is strongly osteogenic in cultures of isolated bone marrow stem cells (data that we have replicated). Vignery and coworkers have demonstrated that when nominal quantities are injected into rat, KGP prevents ovariectomy-induced decreases in tibial bone volume through stimulating osteoblastic action (substantial increases in the osteoblastic surface, osteoid volume, and mineralization rate).

Our initial studies support a biological role for KGP in skeleton by demonstrating rapid and direct cellular actions on cultured osteosarcoma and osteoblastic cells that appear independent of adenylate cyclase activation. Our data suggest that KGP coactivates 3 separate signaling pathways in osteoblasts, most notably novel activation of membrane ATP-dependent K+ (Katp) channels and resultant attenuation of calcium ion-uptake. Transduction

Figure 5.8 *(continued)*

of the KGP signal rapidly results in specific changes in cellular levels of mRNAs encoding both metabolic/signaling and phenotypic osteoblastic proteins important to bone formation. We have now focused on testing whether Katp channel activation or modulation of cytosolic [calcium ion] is responsible for KGP's stimulation of bone sialoprotein (BSP) and procollagen I(alpha) (Col I) gene expression, markers of osteoblastic function. Neither CT nor forskolin replicate this stimulation of BSP and Col I gene expression, but this effect is mimicked by the antihypertensive drug pinacidil, a specific activator of Katp channels.

Based on our initial studies, we postulate that: (1) KGP's activation of Katp channels is a direct action central to all its osteoblastic effects; (2) KGP-induced stimulation of BSP and Col I gene expression is mediated either (a) by activation of membrane Katp channels and attenuation of cellular calcium ion-uptake, or (b) by release of intracellular calcium ion.

To test these hypotheses, we will use rat neonatal osteoblast cultures, and: (a) analyze KGP-induced changes in cellular calcium ion and K+ utilization and membrane potential via fluorescent single-cell imaging and by pharmacological/Rb+1-flux studies; (b) correlate KGP's stimulation of BSP and Col I gene expression with 3 distinct and specific cellular effects produced by KGPs in osteoblasts.

This basic work has direct clinical relevance. Treatment and management of osteoporosis, a debilitating and morbid disease of both aged men and women, is a major healthcare problem, whose estimated 1992 annual cost was in excess of $12 billion. Thirty-three percent of all hip fractures, 40% of forearm fractures, and 13% of vertebral fractures are suffered by males. With an increasingly elderly average population, the incidence of age-related forms of osteoporosis is progressively increasing; it is now estimated that the lifetime hip fracture risk is 15-30% for women and 7-10% for men.

It is difficult to effectively treat osteoporosis when the root cause as well as many aspects of skeletal pathophysiology remain unknown. When the complexities of normal osteoblastic cell regulation are elucidated, it will then be possible to design better therapies. These proposed studies focus on novel regulation of osteoblasts by a skeletal neuropeptide, but they will also broaden our understanding of normal osteoblast biology and cell physiology. Since a key defect underlying metabolic bone diseases such as osteoporosis appears to be a defect in recruitment of mature osteoblastic cells to bone resorption sites, assessment of osteoblastic regulatory peptides may lead to therapeutic peptide development or to novel peptidomimetic therapies.

Source: Section of Formal Proposal by Dr. Douglas Burns, 1997

Figure 5.9   Researcher's Bibliography (Partial)

1. Burns DM, Rodi CM, Agris PF: The natural occurrence of an inhibitor of cell growth in normal and tumorigenic cell lines. *Cancer–Biochemistry and Biophysics* 1:269-277, 1976

2. Burns DM, Touster O: Purification and characterization of rat liver glucosidase II. *Journal of Biological Chemistry* 257:9991-10001, 1982

3. Miller RE, Pope SR, DeWille JD, Burns DM: Hydrocortisone increases and insulin decreases the synthesis of glutamine synthetase in 3T3-L1 adipocytes. *The Journal of Biological Chemistry* 258:5405-5413, 1983

4. Birnbaum RS, Mahoney WC, Burns DM, O'Neil JA, Miller RE, Roos BA: Identification of procalcitonin in rat medullary thyroid carcinoma cells. *The Journal of Biological Chemistry* 259:2870-2874, 1984

5. Miller RE, Burns DM: Regulation of glutamine synthetase in 3T3-L1 adipocytes by insulin, hydrocortisone and cAMP. *Current Topics in Cellular Regulation* 26:65-78, 1985

6. Burns DM, Bhandari B, Short JM, Sanders PG, Wilson RH, Miller RE: Selection of a rat glutamine synthetase cDNA clone. *Biochemical Biophysical Research Communications* 134:146-151, 1986

7. Bhandari B, Burns DM, Hoffman RD, Miller RE: Glutamine synthetase mRNA in cultured 3T3-L1 adipocytes: Complexity, content, and hormonal regulation. *Molecular and Cellular Endocrinology* 47:49-57, 1986

8. Miller RE, Burns DM, Bhandari B: Hormonal regulation of glutamine synthetase in 3T3-L1 adipocytes. *Biology of the Adiposyte: Research Approaches,* G. Hausman, R. Martin, eds, Van Nordstrand Reinhold Co., New York, 1987, pp 198-228

9. Burns DM, Birnbaum RS, Roos BA: A neuroendocrine peptide derived from the amino-terminal half of rat procalcitonin. *Molecular Endocrinology* 3:140, 1989

10. Burns DM, Forstrom JE, Friday KE, Howard GA, Roos BA: Procalcitonin's amino-terminal cleavage peptide (N-procalcitonin) is a bone-cell mitogen. *Proceedings of the National Academy of Science (USA)* 86:9519-9523, 1989

11. Howard GA, Liu C, C. Burns DM, Roos VA: In vivo bone metabolism effects of N-proCT, a novel peptide from the calcitonin gene. *Fundamentals of Bone Growth: Methodology and Applications,* Dixon AD, Sarnat BG, eds. CRC Press, Boca Raton, FL, 345-351, 1991

Figure 5.9 *(continued)*

12. Burns DM, Hill EL, Edwards MW, Forstrom JW, Liu CC, Howard GA, Roos BA: Complementary Anabolic Skeletal Action of CT Gene Products. *Osteoporosis 1990*, C. Christiansen and K. Overguard, eds., Osteopress ApS, Copenhagen, 1301-1307, 1991

13. Edwards MW, Forstrom JW, Burns DM, Roos VA, Howard GA: N-procalcitonin: Kinetics and biodistribution in intact mice, and effect on cortical bone and bone resorption in oophorectomized mice. *Osteoporosis 1990*. C. Christiansen and K. Overguard, eds., Osteopress ApS, Copenhagen, 396-399, 1991

14. Burns DM, Howard GA, Roos BA: An assessment of the anabolic skeletal actions of the amino-terminal peptides from the precursors for calcitonin and calcitonin gene-related peptide. *Annals of the New York Academy of Science* 657:50-62, 1992

15. Kawase T, Howard GA, Roos VA, Burns DM: Diverse actions of calcitonin gene-related peptide on intracellular free Ca-2+ concentration in UMR-106 osteoblast-like cells. *Bone,* 16:379S-384S, 1995

16. Kawase T, Orikasa M, Ogata S, Burns DM: Protein tyrosine phosphorylation induced by epidermal growth factor and insulin-like growth factor-I in the rat clonal RDP-4.1 dental pulp-cell line. *Arch Oral Biol* 40:921-929, 1995

17. Kawase T, Ogata S, Orikasa M, Burns DM: 1,25-Dihydroxylvitamin D3 promotes prostaglandin E–induced differentiation of HL-60 human premyelocytic cells. *Calcified Tissue International* 57:359-366, 1995

18. Kawase T, Howard GA, Roos BA, Burns DM: Calcitonin gene-related peptide inhibits net transmembrane Ca-2+ uptake in osteoblastic cells through cAMP-independent activation of ATP-sensitive membrane K+ channels. *Endocrinology* 137: 984-990, 1996

19. Kawase T, Oguro A, Orikasa M, Burns DM: Characteristics of NaF-induced differentiation of HL-60 cells. *J Bone Min Res*, 11:1676-1687, 1996

20. Kawase T, Howard GA, Roos BA, Burns DM: Acute inhibition of Ca-2+ uptake in osteoblastic UMR-106 cells by parathyroid hormone and prostaglandin E2: Comparison to the effects of calcitonin gene-related peptide. *Endocrinology,* 1996

21. Kawase T, Howard GA, Roos BA, Burns DM: Nitric oxide stimulates osteoblast-mediated in vitro mineralization. *Bone,* 1996

Source: Section of Formal Proposal by Dr. Douglas Burns, 1997

Figure 5.10 Table of Abbreviations

Abbreviations Used in This Narrative

KGP—the neuropeptide "kalrectin gene-related peptide"

CT—the systemic hormone calcitonin

PTH—parathyroid hormone

VIP—vasoactive intestinal peptide

NPY—neuropeptide Y

K_{atp}—ATP-sensitive potassium channel(s)

KCa—Ca2+-dependent potassium channels

Pin—pinacidil, a specific activator of Katp

Gly—glyburide (glybenclamide), a specific activator of Katp

TEA—tetraethylammonium, a selective inhibitor of K_{ca}

PKC—protein kinase C (Ca^{2+} – dependent protein kinase)

PMA—phorbol 12-myristate 13-acetate, an activator of many PKC isoforms

VDC channels—voltage-dependent Ca2+ membrane channels

Diltiazem—specific inhibitor of L-type VDC channels

bis-oxonol—the potential sensitive dye, (bis-(1,3-diethylthiobarbituric acid)-trimethine oxonol)

TFA—trifluoroacetic acid

FK—forskolin, an activator of adenyl cyclase

[Ca2+]i—concentration of intracellular calcium ion

BSP—bone sialoprotein

Col I—short for the alpha(1) chain of collagen I

OPN—osteopontin

ON—osteonectin

iNOS—inducible nitric oxide synthase

GS—glutamine synthetase

AR-S—the calcium-specific dye Alizarin Red-S, used in biomineralization assays

TGF-beta—transforming growth factor type beta

IBMX—isomethylbutylxanthin, an inhibitor of cellular phosphodiesterases

Obch—porin-like osteoblastic channel protein

GAPDH—glyceraldehyde 3-phosphate dehydrogenase

A1—nonregulated constitutive mitochondrial protein of unknown function

8-BcA—8-bromo-cAMP, a stable cAMP analog

BAPTA—short name for a complex compound which acts as a Ca2+ sponge

PSS—physiologic sterile saline solution

TCM—tissue culture medium

Source: Section of Formal Proposal by Dr. Douglas Burns, 1997

Figure 5.11   A Graph Helps Readers Understand the Time Line of Experiments Proposed in the Report

Activity	Year 1	Year 2	Year 3	Near Future
A. Single-cell imaging studies				
Study of cell Ca^{2+}	██████████	████		
Study of cell K^+	██████████████			
Bis-oxanol assay of Em hyperpol.	██████████	██		
B. mRNA study & correlation of regulation with known KGP signaling				
Cellular levels dose-response time-course	████████	████		
Test effect of cellular K^+ and Em hyperpolar		████	████	
Test effect of cellular Ca^{2+}		████	██	
Retest effects of cellular cAMP		████	██	
Extend mRNA findings to study of transcriptional activation				████████

Source: Section of Formal Proposal by Dr. Douglas Burns, 1997

Figure 5.12   A Graph Helps Readers Visualize Material in the Main Part of the Report

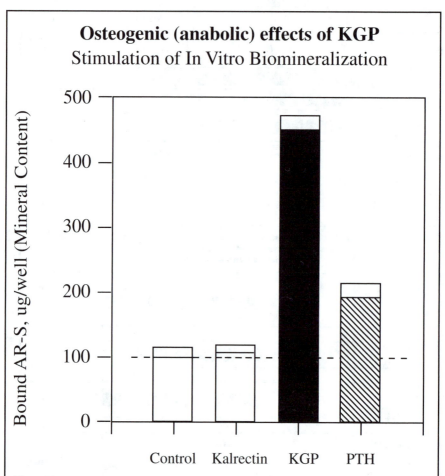

Osteogenic (anabolic) effects of KGP
Stimulation of In Vitro Biomineralization

Fig. 5: KGP stimulates mineralization in osteoblastic cells.

For these quantitative assays, cells cultured in BGJb media supplemented with 1% fbs were treated for 25 days with the indicated biological agent (media changed every 4 days). At the end of day 25, cells were fixed with 70% ice-cold ethanol (60 min). The ethanol is aspirated, the cells are rehydrated with water (10 min), and then the cultures are stained with 40 mM Alizarin Red-S (AR-S) for 10 min. Cultures are washed extensively with PBS to remove nonspecifically bound stain, and then bound dye is solubilized with cetylpridium chloride (10% in 10 mM sodium phosphate, pH 7.0). Aliquots and dilutions are read in spectrophotometer and the absorbance at 562 nM read against an AR-S absorbance standard curve for quantification. The bars indicate the total amount of AR-S specifically bound to the calcified extracellular matrix in these cultures (mean ±SD; n = 3 dishes per treatment).

Source: Section of Formal Proposal by Dr. Douglas Burns, 1997

How It Works. The main units of this multisectioned report include:

◆ *A cover sheet* acting as both transmittal describing the purpose of the document and as its title page, containing contact information and the project's working title

◆ A *detailed abstract* (Figure 5.8) explaining the goals of the research, and serving as an "executive summary"—quite a lengthy one by business standards, but acceptable in the realm of sophisticated research

◆ *Funding justifications* showing the budget for equipment, staff, and overhead

◆ A *biography* of the researcher, also containing a bibliography of the researcher's publications (Figure 5.9), that acts as a references or credentials section

◆ A *table of abbreviations* used in the study (Figure 5.10) so that reviewers can understand the narrative in the application, or have key terms defined for them

◆ A *reply to preliminary reviewers* of the project, necessary to show that the project design is valid and overcomes criticisms

◆ The *proposal itself*, which divides into:
> *Rationale* for the project
> *Discussion* of research already accomplished (initially summarized in Figure 5.8)
> *Detailed explanation* of experimental design (see Figures 5.11 and 5.12)
> *Bibliography* for the whole project
> *Back matter* (certifications of proper protocol and safety; letters of reference)

Notice that this outline reflects the integration of a proposal or prospectus within the larger framework of a major grant application. Writing which validates the authenticity of the writer and subject precedes the proposal itself.

Although these documents do not necessarily conform to the expected template of the simple proposal, they build upon that template nevertheless, even as they adhere to the formal requirements, organizational wishes, and goals of the recipient. There is also a basic message—support—closure format operating in most such proposals, which are engineered to follow the elements of good technical writing.

Advice

Considerations to Remember in Writing Grant Proposals. I call these the "Six Fs" that you must contemplate as you draft your proposal.

◆ _Field_—Why is this project important in its field? You must explain what differentiates your work from that of others. Are you building upon ideas already conceived, but taking them in new directions? Are you synthesizing the seemingly opposing viewpoints of others? Is your project a radical rejection of the work of the past?

◆ _Function_—What are the project's tasks, goals, detailed objectives? Proposals will ask you to explain these fully. Unless you can explain them to yourself, on paper, you will have difficulty convincing others of their validity.

◆ _Framework_—What are the project's constraints? How big is it? What is its scope? Outside of what parameters would your project not yield valid results?

◆ _Fallout_—What are the benefits and detriments if the project is implemented? Are there anticipated results that could help us see the issues covered in a new way?

◆ _Format_—What are the specific requirements of form? What presentation standards are current in my field? What does the grantor want? (These last two questions are sometimes at odds.)

◆ _Finance_—How will the project be funded? Will there be a matching contribution from me or from my institution? Am I seeking secondary support from another grantor?

These questions, crucial to the success of your project and proposal, appear in Figure 5.13 in worksheet form for your convenience in early planning.

Figure 5.13   Worksheet for Thinking about Proposed Projects

Project Worksheet

(Ask Yourself These Questions)

Field:
What is the primary importance of the project?
What secondary benefits are there?

Function:
What are the goals or objectives?
Can I define the tasks to do?

Framework:
What is the project's role and scope?
Will there be limitations on its success?

Fallout:
Can I define the project's positive implications?
What are its negative implications?
Will there be new results applicable in a new way?

Format:
Am I writing in an accepted style?
Am I using the correct method of documentation?
Are there unusual requirements to be followed?

Finance:
What are the sources of funds?
When will the funds be released?
What is the schedule of expenditures?
How will I display the budget on paper?

Patterning in Grant Proposals.   Certainly grant proposals can contain such items as abstracts, reviews of literature, method and reference sections that are also aspects of the formal proposals we have seen in earlier chapters, but these may occupy different places in the sequence than would be expected in other proposal writing. For example, a proposal for a large science grant may contain

◆ A *cover sheet* that is itself a form produced by the grantor and that is originally provided in an application packet
◆ A *project summary* or abstract
◆ A *table of contents* often printed on a special form supplied with the application

- A *description of the proposed project and of past related projects*, if funded by the grantor
- *References* used in the proposal
- *Biographies* of the investigators, experimenters, staff
- *Budget statements* showing in detail the expenditures by year and in total, and containing a justification of such expenses
- An *explanation of any current funding or known future funding* for the project
- A *description of all equipment, materials, sites* to be used
- *Other relevant items*
- *Optional appendix*

Or, some grant proposals may look like this:

- *Cover sheet.* Often supplied in the application packet as a form to which you add your specific information, this sheet may also include the amount of funds sought, the time the project will take, the proposal writer's contact information, and the signatures of approving officials at the writer's institution.
- *Abstract or summary.* This section must be designed to provide a condensed and clear description of the project as if it alone will be read— sometimes that is the case in the early reviewing of proposals. Sometimes the abstract appears on the cover sheet.
- *Explanation of mission or purpose of the project.* This must explain how the project responds to specific needs generated by a problem. Sometimes the mission statement may be phrased in terms of specific outcomes. Other ways of stating the mission include a discussion of the hypothesis or the presenting of research questions requiring answers.
- *Rationale.* This section states concisely why the project should be done. It might be combined with the previous part of the proposal. Sometimes a literature review is combined with the rationale.
- *Methods.* This describes the procedures of the project, including what will be done and when, who will perform each task, how the project will be executed, and what may result.
- *Evaluation/assessment of project.* This part may be addressed in "Methods," but if not it can stand as a separate section. Readers will need to know if there is a single, final accomplishment of the project that will be assessed, and if so, how—or if the operations leading to the accomplishment are each to be evaluated, and if so, how. Often grantors have specific forms and procedures for such evaluations, including questionnaires to use when giving public presentations. In fact, sometimes the evaluation section is followed by one that profiles the audience of the project.
- *Credentials.* This segment describes the credentials of those seeking the grant. These include the project director and participants as well as the institution or agency. Appearing in this part might be a bibliography of the works by the principal researcher or director, curriculum vitae, lists of references. Perhaps the institution's certifications of safety and ethical compliance will appear here; if not, such will be found in "supporting documents."

◆ *Response to reviews.*   If there have been preliminary reviews of an earlier version of the proposal, this section may appear in the prospectus. Its purpose is to emphasize the strengths and to respond to any perceived weaknesses in the design of the intended project.

◆ *Finances/budget.*   Sometimes this section appears as a set of budget worksheets in an appendix.

◆ *Supporting documents.*   This part is often the location of supplementary references, institutional research approval, and certifications of compliance.

Each application is unique. Such categories as those just mentioned are common to most documents but may vary in name, sequence, or inclusion. Recall their corresponding parts in some academic or business proposals, though (see Chapter Four), and you will see a close connection, since in both cases the disbursing entity wants to be sure the money will be well-spent:

Prospectus for an Academic Project	Business or Institutional Proposal
Abstract (if required)	Executive summary
Introduction	Introduction
Problem	Statement of challenge
Hypothesis	Statement of agenda
Project method	Means to address challenge
Expectations	Expected results
Timetable	Timetable
Appendix	Supporting material
References	References

For Further Study

1. How do grant proposals reflect the style and strategy of shorter prospectuses?

2. If you were to produce a multimedia presentation to accompany a grant application, what would determine your selection of content and means of display?

3. If you gave an oral presentation of your application using multimedia to assist you, what challenges would you face in integrating into it the content of the written document and of the computer presentation?

Review of Patterns

General Template of Many Grant Proposals

Cover sheet

Abstract or summary

Explanation of mission or purpose

Rationale

Literature review

Methods

Evaluation/assessment

Credentials

Response to reviews

Finances/budget

Supporting documents

Or,

A *cover sheet* that is itself a form produced by the grantor and that is originally provided in an application packet

Project summary

Table of contents

Description of the proposed project

References

Biographies

Budget

Current funding

Equipment, materials, sites

Other items

Appendix

An Extended Grant Proposal

Main document

- Cover sheet

- Abstract

- Funding justifications

- Biography of the researcher

- Table of abbreviations

- Reply to preliminary reviewers

- Proposal itself

  - Rationale

  - Discussion of research already accomplished

  - Experimental design

  - Bibliography

  - Back matter

References

Now that you have surveyed the writing of proposals, the best way to become acquainted with such documents in your field is to study professional journals and the Internet for proposal requests, obtain applications, and work through the challenges involved in crafting these documents yourself. Following are some helpful general resources to consult as you develop your prospectus.

American Psychological Association. *Publication Manual of the American Psychological Association*. 5th ed. Washington, DC: APA, 2001.

This is the standard manual for APA citation and format in the social sciences. The book explains APA style in detail and provides models for use. APA also offers a supplemental disk to assist the writer.

Borowick, Jerome. *Technical Communication and Its Applications*. Englewood Cliffs, New Jersey: Prentice Hall, 1996.

This book contains material on proposals and feasibility reports, emphasizing clean presentation in its sample documents.

Gibaldi, Joseph. MLA *Handbook for Writers of Research Papers*. 5th ed. New York: Modern Language Association, 1999.

This is the standard exposition of MLA formatting and style. MLA style is employed often in work discussing the arts and humanities. Even if that is your style of choice, you might benefit from consulting Turabian, listed later, for clear explanations about such things as tables and vertical lists that might be part of your document.

Hall, Mary S. *Getting Funded: A Complete Guide to Proposal Writing*. 3rd ed. Portland, Oregon: Continuing Education Publications, 1988.

A guide for grant writers, though not a guide to proposal writing in general.

Holloway, Brian R. *Technical Writing Basics: A Guide to Style and Form*. 2nd ed. Upper Saddle River, New Jersey: Prentice Hall. 2001.

A general survey of technical writing with material on proposals and related documents.

Lannon, John M. *Technical Communication.* 8th ed. New York: Longman, 2000.

A large, workplace-oriented text with some general material on proposals.

Locke, Lawrence F., Waneen Wyrick Spirduso, Stephen J. Silverman. *Proposals that Work: A Guide for Planning Dissertations and Grant Proposals.* 3rd ed. Newbury Park, California: Sage, 1993.

A helpful book for the graduate student, particularly the student working with a project in social sciences or education.

Pearsall, Thomas E. *The Elements of Technical Writing.* 2nd ed. The Elements of Composition Series. Boston, Massachusetts: Allyn and Bacon, 2001.

An overview of technical writing including a general view of proposals.

Pfeiffer, William S. *Technical Writing: A Practical Approach.* 4th ed. Upper Saddle River, New Jersey: Prentice Hall, 2000.

Emphasizes the corporate realm of technical writing, including proposals.

Tornquist, Elizabeth M. *From Proposal to Publication: An Informal Guide to Writing about Nursing Research.* Menlo Park, California: Addison-Wesley, 1986.

Designed for nursing students, but useful to others embarking on extended projects. It contains some helpful suggestions about organizing proposal material conceptually.

Turabian, Kate L., John Grossman, and Alice Bennett. *A Manual for Writers of Term Papers, Theses, and Dissertations.* 6th ed. Chicago: U of Chicago P, 1996.

Presenting the essence of Chicago style, this book assists readers with issues of tone and format as well—whether or not these readers will be using this style.

White, Virginia. *Grant Proposals that Succeeded.* New York: Plenum, 1984.

A handbook for grant writers.

APPENDIX

A Review of Technical Writing

Expectations

Like other types of writing that address specialized needs, technical writing has its own requirements of style, audience, tone, structure, format, and documentation. However, the specific expectations that your document must fulfill will vary from discipline to discipline. For example, legal documents are commonly double-spaced, whereas many business proposals employ single-spacing. Some grant proposals use centered headings. Some use flush-left headings. A proposal's preliminary abstract may appear inside the document or on its cover sheet. Despite such variety, we can make certain generalizations about the appearance, format, and goals of technical writing.

Style

The goal of technical writing is to present information effectively on the page for maximum results, whether those results are to inform or to persuade. Such writing resembles the academic essays assigned in college English courses in its attention to unity, support, and coherent style. Both types of communication require clarity, focus, audience-awareness, development, coherence, and smooth expression—the absence of problems with word choice and grammar.

But the technical document's goal of achieving a targeted response transcends that of achieving a good grade from an instructor whose primary interests might well be the elegant writing of the academy, with its absence of headers, generally lengthy paragraphs, and often complicated sentences. The technical writer must use exact words in clear syntax to promote, for example

◆ the understanding of information
◆ the acceptance of a proposal
◆ the consideration of a plan
◆ and the establishment of goodwill

Diction.   Your choice of words is as important as adhering to the rules of grammar. A bossy tone alienates readers; the irrelevant use of specialized terminology may antagonize nonspecialists; clichés prevent understanding by filling up space with meaningless phrases; inappropriate word choice, including unconscious punning, deflects attention from the issue presented and may offend. Watch out for these problems, discussed next, which may creep into the best of proposals.

Authoritarian language.   You can recognize authoritarian language in such phrases as "It has come to my attention that . . ." Note that this example focuses on the sender of the message rather than the recipient. Yet it is the recipient whom you must convince to act upon or to accept your information. Technical writing adopts the attitude that communication exists to benefit the person receiving the message.

Passive voice.   Unnecessary use of the passive voice, in which the subject of the clause becomes the object and the object the subject, may also sound pompous or obscure, as it removes the "doer" of the action from its normal word order. Avoid "The marketing campaign was implemented by the Lyons Corporation" but say instead "The Lyons Corporation implemented the marketing campaign." Reports and proposals will contain a certain amount of the passive, often to avoid egotistical use of "I," but try to eliminate passive constructions where possible. Not "The project is divided into two phases," but "The project divides into two phases." Not "The plan was executed," but "We executed the plan."

Jargon.   Misapplied technical language, or jargon, occurs when overly-specialized language attempts to communicate with a general readership or when the language of one field is used inappropriately in another. Some jargon words have become so worn with overuse that they possess at best a diluted meaning. For example, a "venue," originally the locale of a judicial proceeding, now means any location at all. Thus today a beautician, a plumber, a film, and a salamander might all have "venues." The word "hypertext" has a specific reference within computer science; however, it is also likely that any utterance longer than one sentence may be deemed "hypertextual" in its current popular meaning of "a bunch of different things all stuck together." And examine the bureaucratese that follows, noticing that you may combine the words in Column One with those in Column Two in any way whatsoever, before adding the resulting phrase to the hapless noun of Column Three, all without establishing the meaning of the aggregated expression:

Column One	Column Two	Column Three
maximal	environmental	friendliness
differing	functional	orientation
viable	programmatic	implementation
growth-driven	enhanced	development

Since there are no clear definitions for these statements, they seem designed to be inserted into bureaucratic documents at will. In fact, several entertaining sites on the Internet create, at random, nonsense "essays" composed of jargon-words at the touch of your keystroke. Remember that if you can't picture the meaning of a phrase clearly in your mind, you must change the phrase until you can. Think of the picture first, and then create the phrase to describe it—that's what George Orwell advises in "Politics and the English Language," an essay still relevant for our imprecise times.

Clichés. Clichés are commonly-employed fillers in conversation. As does jargon-choked language, cliché-ridden text lacks meaning. *It's best to avoid clichés like the plague, so as not to throw the baby out with the bath water and leap from the frying pan into the fire. Y'know?* Again, visualize what you want to say, then say it precisely. If a cliché does happen to work in context, change it anyway, using distinctive synonyms.

Inappropriate language. Manifestations of this problem include subconscious wordplay, such as inadvertently asserting that a sewer tax won't drain one's finances. Sexist phrasing is also a species of inappropriate language: a memo referring to all executives as exclusively of one gender is sure to provoke irritation. English paradigms contain several options for dual-gender and gender-neutral writing. For instance, you may

- Use plural forms to include both genders.
- Use the neutral "one" in place of gender-specific pronouns.
- Rewrite the text to eliminate gender. This strategy avoids the rough reading which another choice, the split pronoun, frequently causes: "Those are the options available to him/her" is considerably more awkward than "Those are the options available."

Of course, no racial, ethnic, or sexual derogation should ever appear in your writing.

International Concerns. English is a world language, so removing local phrases from international correspondence becomes critical. Such limited expressions as clichés, references to stateside collegiate athletics, and the illogical utterances of idiom can make no sense to readers for whom English is not the primary language. "World" or "International" English, a trade language and the discourse of computer information, must remain free of localized encumbrances to work.

Repetition is a type of poor word choice that usually indicates failure to combine thoughts into complex or compound sentences or to use parallelism. It lends to a statement the effect of being stuck in reverse. For example: "In deed, if not in action; in thought if not in belief, in words if not in phrasing; systemically, if not internally, it inheres."

Grammar itself. No one is immune from mistakes in grammar and usage. You should have a kit of standard writing equipment including a good dictionary such as *Webster's* or *American Heritage*, a current manual of style and grammar such as the *Prentice Hall Reference Guide to Grammar and Usage* by Muriel Harris, and a thesaurus, or dictionary of synonyms. Be wary of the "thesaurus" function in your computer program, since it seldom provides the nuances that distinguish the implied meanings of words. Also, spelling-checker or grammar-checker software, though exceedingly helpful, will not provide certainty and accuracy.

Audience

The communication in a proposal might best be described as "transactional." In a transaction, there are two parties. One proposes something to the other, but by accepting the ideas in that document the other party may have to surrender something—money, time, even beliefs and values. The other party may have to devote valuable moments just to *reading* the proposal. The strategy of persuasion underpins this form of communication, since it is always difficult to convince people to act.

In order to make a written transaction successful, you must know the audience. Research your audience carefully. What are the audience's needs, values, goals? You would not, for example, send a proposal to fund a new work of art to a humanities foundation unconcerned with the primary production of new work. Also, consider the comfort of your audience. No executive is going to have the time to wade through a densely-typed, leisurely essay dawdling across four pages unrelieved by white space, decoding the multiple subtexts of the intricate communiqué by seeking its deconstructive sutures. Instead, clarity is paramount. This puts the burden on the writer, who must direct the reader, using headings to show the outline of the discussion, and lists to clarify the points discussed. Such an approach encourages the reader to follow the document as it proceeds.

Tone

Knowing exactly what you want your audience to do will help you convey your message with the proper tone. Remember that the writer of a proposal is asking for something which may not have to be granted. Avoid giving the impression of arrogance or of servility. Include neither patronizing nor subservient diction, such as the bossy "As you will probably understand if you read Section Three of the proposal . . .," or the whiny "Our very hard work on this project, especially shown in Part Two . . ."

Content

You must present clear and specific content within business and technical format.

Working Together to Develop Content.   One of the biggest challenges in business and technical writing is that frequently the different sections of a document are contributed by different people, each having particular stylistic quirks, and each emphasizing some things that may not be important to the document as a whole. Then the group must decide how to reconcile all the parts with each other, what to enhance, what to discard, and how to integrate parts so that they become a seamless unity.

Continuity. And in group situations, documents are often "grown" from small, isolated sections into developed, multi-unit presentations: major proposals, employee handbooks, feasibility reports. Much give-and-take and many hours of re-examining the drafts produce a finished text. Many writers begin with an outline which shows the major headings in place; the supporting material under each heading is developed separately by different people; then the whole package is put together, reviewed, and re-reviewed. What looks like a natural, effortlessly-produced presentation is really the product of intensive work.

Figure A.1   A Process of Developing Content

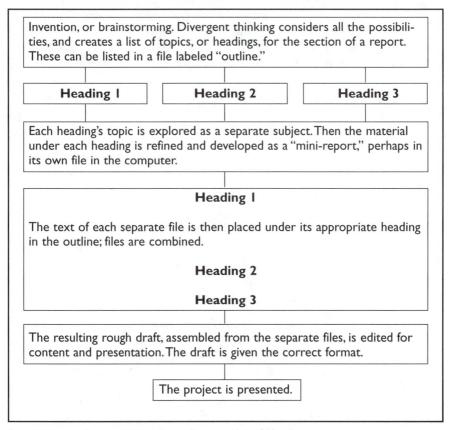

Invention, or brainstorming. Divergent thinking considers all the possibilities, and creates a list of topics, or headings, for the section of a report. These can be listed in a file labeled "outline."

Heading 1	**Heading 2**	**Heading 3**

Each heading's topic is explored as a separate subject. Then the material under each heading is refined and developed as a "mini-report," perhaps in its own file in the computer.

Heading 1

The text of each separate file is then placed under its appropriate heading in the outline; files are combined.

Heading 2

Heading 3

The resulting rough draft, assembled from the separate files, is edited for content and presentation. The draft is given the correct format.

The project is presented.

From Holloway, *Technical Writing Basics,* Prentice Hall, 1999, p.5.

Using Computers. Even writing by yourself, you'll undergo similar challenges. It's easiest to develop a document on disk, using a computer program
with which you are familiar. Big projects often begin with material in separate files, which can, once developed, be imported into the main shell of the
document. That shell should be an outline consisting of headings and sub-
headings. In fact, the best way to organize small-to-mid-size projects is to determine the pattern of organization, express the pattern in an outline form
displaying key words or phrases, and plan on those key words serving as
main and sub-headings for the document (see Figure A.1). Having such an
outline, or shell, (as shown in Figure A.2) means that each paragraph or paragraph sequence governed by a heading can be developed separately and
then inserted into the appropriate place in the document. In Figure A.2 a
shell for the research proposal exists, assigned by the instructor. The student
creating such a document need only respond to the prompts in the premade
outline. Moreover, material under each heading can be developed by itself,
then integrated into the forthcoming master document later. The proposal,
once refined and approved, will be the basis for a longer researched report.

There are other good reasons to use computers in drafting your work.
They make proofreading easier, since you don't need to waste paper, since
the text looks clear, and since programs can check your spelling. They permit
returning repeatedly to the document and tinkering with it. They promote
preparing alternative versions of the same project so that you can pick the
best one. And they provide the best options for attractive display of material.

Figure A.2   Use of "Shell" Pattern in Proposal Writing

Header
*(Use standard
heading of To, From,
Date, Re.)*

Title
(Should be the title of your project)

I. Summary of Topic
*(First subheading
will also become the topic of your final paper)*

*An explanation in paragraph form underneath this heading would summarize
the topic, lead into the thesis, and present the thesis. The thesis of this paper
might be a problem-solution one—taking the general form of "claims of fact plus
claims of value produce claims of policy"; that is, because of the problem you've
just summarized, and because we value certain things, we must implement a particular solution. You might discuss a tentative thesis as an hypothesis subject to
change when you know more data.*

Figure A.2   *(continued)*

II. Development of the Paper

In this section, you'd explain how the body of the paper would be organized. The BEST way to do this is to begin with a statement saying "At this time, I intend to organize my paper to answer the following questions." Then, state the questions, explaining in short paragraphs the significance of each:

Question 1: What is _____? (Definition).

Question 2: How did _____ become a problem? (Background).

Question 3: What are the specifics of the problem?

Question 4: What solutions have been proposed for this problem?

Question 5: Which is the best solution, and why?

These questions define the basic concerns of a problem-solution research paper.

III. Methods of Research

Begin with a paragraph or two discussing your research strategy—covering what material you've looked at, and the types of material you have yet to use. Will the bulk of your research come from books, articles, the Internet? What conventional and electronic sources work best for this topic?

Follow up with a Works Cited list in standard MLA form (or current APA form, if you intend to do your paper in that style). A research proposal usually lists a minimum of five to seven sources.

IV. Timetable

(In this section, explain how long the phases of the project will be, and when the project will be done)

V. Request for Approval

(Ask that the project be approved, and request any guidance deemed appropriate)

Doing your homework and convincing the reader with detail is only half the challenge. You must also adhere to the page layout and style of presentation expected in order to get results.

Structure

The technical communicator should become familiar with the types of order commonly used to present information (see Figure A.3). A proposal employs

Figure A.3 Types of Order Located in a Proposal

To:
From:
Date:
Re:

Title

I. Summary of Topic

A beginning discussion section presents a SUMMARY overview of the proposal that may include an initial example leading into a thesis or statement of purpose—a PERSUASIVE strategy.

II. Development of the Paper

Because this section explains how the body of the paper will be organized, it is ANALYZING the material.

Question 1: What is _____? (SUMMARY).

Question 2: How did _____ become a problem? (PROCESS).

Question 3: What are the specifics of the problem? (ANALYSIS).

Question 4: What solutions have been proposed? (DESCRIPTION).

Question 5: Which is the best solution, and why? (COMPARISON and PERSUASION).

III. Methods of Research

First ANALYZES research strategy—covering material looked at, and the types of material yet to be used.

Follows up with a reference list providing a SUMMARY . . . .

different kinds of structure in its different sections. The use and the recognition of patterns create psychologically-satisfying writing which adheres to the reader's expectations. The patterns in business and technical writing are few but powerful. Let's examine them. Figure A.3 is a partial sketch of the student proposal depicted in Figure A.2, but with its sections labeled to show the kinds of order used.

Types of Order. The everyday world contains so much information that any transactional writer must select from that vast storehouse to present a limited amount of data to the reader. From the chaotic world, writers mold material into logical presentation. That can be difficult. For example, if you have been angered by rude or incompetent service at a business and have tried immediately afterwards to compose a letter to the management, your anger may prevent you from being selective. The onrush of emotion may cause you to write associatively rather than logically. But the reader of such a letter will not be inclined to help you, or indeed, to finish the letter, unless all that anger is contained and material is selected to present in logical order. Such logical order has three main parts. It typically begins with a statement of claim, or a question to be answered. Claims may stress factuality, emphasizing that a certain event happened. Some claims insist on preference. Many claims urge action. Once the claim is established, transactional writing then discusses details. Transactional or business writing then closes with a statement that encourages an obligation.

Essentials of Order. As you have just seen, logical order involves a pattern similar to the thesis-development-conclusion approach you learned in English composition. Putting the thesis at the beginning of the document results in a *deductive* approach which declares the point in the introduction, supports it in the body, and clinches it in the conclusion. Beginning with a question, or a hypothesis, following with data which must be understood, and concluding with an answering thesis constitutes the *inductive* approach. At its most basic, the support in the body might be description (discussion of appearance, geometrical arrangement in space, texture, or hue) or narration, which depicts an event moving through time.

There are differences, however, between basic essay writing and the transactional writing present in proposals. Transactional writing may not be unified by a thesis such as "We need a campus day-care center" at all, but rather by a statement of purpose, as in "This document will demonstrate the need for a campus day-care center." Often the thesis and purpose statement can be combined: "Because of x, y, and z, we need a campus day-care center. This document will discuss the ways in which we can achieve this goal."

Specific Templates. Governing the general logical pattern often used in proposals—which we can call

◆ Message
◆ Support
◆ Closure—

will be a specific template best suited for the job. Such templates include:

◆ *Summary* (condenses a description of something, frequently narrating to do so).
◆ *Process* (shows how to perform a task or explains a procedure; is based on narrative).
◆ *Analysis* (divides something into its parts to discuss them—perhaps to evaluate them).
◆ *Comparison* (analyzes two or more different things).
◆ *Persuasion* (writes to convince the reader to believe or act).

Summary.   Summary informs readers who need a "quick take" on an issue. The author of a summary must isolate the key elements of the material to be reported and state them in a coherent, organized fashion, using narration, the discussion of events moving through time, and description, which presents spatial material in detail. You'll find summary as a stand-alone entity or as part of a longer document. A summary which precedes and describes a long report or proposal is often called an executive summary or an abstract—it explains the gist of the document so that the reviewer acquires an immediate general understanding, crucial to the routing of the document to those appropriately equipped to read it.

Another common summary is the condensation of a report or document other than the proposal you have written. Such summaries appear in the support sections of proposals. When writing such a condensation, be sure not to duplicate the word order or the language of the original document, and to identify the source of the document (more about this appears under "Documentation," below). Here is an example of a summary of a longer report:

> The Sizemore report surveyed the need for vertical parking space at eight mid-sized hospitals in suburban Chicago. It calculated the square footage required to double the accommodation and analyzed the construction costs per square foot for representative structures. The report also compared the costs and benefits of vertical garages to those of conventional parking lots. Sizemore favored the vertical option.

Process.   Process writing seeks to explain how to perform a task or how something works. It depends on narrative time transitions such as "first," "next," "then," "last." Some process explanations in the workplace are terse operating instructions employing imperative diction and short parallel phrases.

Other types of process writing also use parallel structure and group tasks into steps, but these forms can be more complex. Such types include recipes, how-to books, in-house procedure manuals, sophisticated assembly instructions, and explanations of what happened. You use descriptions of process in your proposal writing (see Figure A.4), since you have to explain how the project will unfold, who will do what, and what time line determines activities. Figure A.4 shows a section from a long document requesting grant funding.

Figure A.4   A Process Section in a Proposal

I shall use the 1996 stipend to travel to the University of Missouri-Columbia, to work with the John Neihardt collections at the Western Historical Manuscript division of the Elmer Ellis Library for four weeks during the summer of 1996. I will research primary materials and artifacts produced by Neihardt's collaboration with Black Elk—a project which resulted in *Black Elk Speaks*. I have already begun this project by contacting the librarian in charge of the Western Historical Manuscript Collection at the Elmer Ellis Library. This collection houses Neihardt's original manuscripts, field notes, supporting materials, and letters. Travel and lodging information is attached in an appendix.

This section provides an overview of what the funds will be used for and when the project will take place.

Analysis.   Process derives from the time sequencing of narrative, but analysis derives from description which doesn't need to be time-dependent at all. Analysis occurs whenever a writer breaks a topic into its components in order to discuss them, or whenever a writer groups related topics under a general heading to explain what they have in common (see Figure A.5). Reviews of cars, stereos, and other merchandise that appear in consumer magazines are examples of analysis; these articles provide a breakdown of an item's features and evaluate the merchandise based on how its parts work. But analysis also applies to the practical issues of your proposal, since you must divide the issue into its separate topics in order to discuss it. Research proposals and many other documents also contain a review of literature that analyzes and evaluates what has been written about your subject. Figure A.5 is an excerpt from a section of a proposal showing the expected benefits of a plan. Analysis is present because the benefits are shown as components of a whole. Each component is discussed separately within a common analytical framework—that of the bulleted list.

Figure A.5   Analysis in a Proposal

Though such a degree is not directly sequential for many teachers, it would provide a good "upgrade" for public school faculty—and others—interested in

- improving Internet capabilities, including computer markup languages and Website creation, as such skills relate to their disciplines;
- understanding literature, culture, and science as expressions of each other;
- enhancing media studies;
- serving special populations.

Comparison. When an analysis involves two different items measured against each other to see which one is better then the analysis is called a comparison. A marketing report might compare two brands of frozen yogurt; a feasibility study might discuss the merits and demerits of two health-care plans; a pre-election prospectus might provide analyses of two or more candidates' views, showing points of agreement and disagreement. There are three types of comparison, and they should not be jumbled together in the same section of a document.

Segmented comparison may compare each sub-system or category of items A and B; for example, the environmental impact of two legislative bills in the first paragraph, the monetary impact in the second paragraph, and the legal impact in the third paragraph. Such comparison discusses a number of small items corresponding in some way. Often in technical writing such as proposals, this comparison appears as a table or chart.

Holistic comparison, used when the units are large and few, first discusses item A in its entirety, then analyzes item B; for example, the original environmental protection plan in a first paragraph; the recent environmental plan in a second paragraph.

Likeness-difference comparison shows how the two items are more alike than different, or more different than alike. Perhaps the two environmental plans are alike in name only. In that case, the comparison would acknowledge that fact and explain their differences.

Persuasion. Ancient rhetoricians contemplated how to keep people's attention during long persuasive speeches. For a presentation to be effective, they said, it must acknowledge the *audience*, *illustrate* the problem to be addressed, and state its *point*. Then it must *clear up* any misunderstandings about key terms or the path the presentation will take. Next, it must *present* the pros and cons of the issue. Finally, the presentation should *conclude* by supporting one position, and asking the audience to do the same (see Figure A.6). Notice that the basic pattern might be outlined like this:

Message
    Address audience
    Illustrate problem
    State point (or thesis)

Support
    Define unclear terms
    Examine pros and cons

Closure
    Affirm position
    Provide closure to discussion.

As does the persuasive speaker, the *persuasive writer* must coach an audience well. In persuasive business and technical documents, the ultimate transactional

Figure A.6 Persuasive Section in a Proposal

Our restoration work is rated nationally as excellent, our timeliness in executing jobs is proven, and our rates are well below the metropolitan average. Please consider how Brand Renovations can benefit you in outsourcing your contractual needs; then phone us at 555-5467 to begin a new and profitable relationship.

---

writing, the writer has to make the work of reading a document seem easy and worthwhile. Instead of vocal transitions and gestures, headers, boldface, and underlined type encourage the reader to keep going; lists bordered by white space urge the subliminal message that "you can understand this: it's easy because it's broken down clearly for you." Proposals must coach the reader, beckoning him or her to follow the sequence of ideas presented. The closing section shown in Figure A.6 recapitulates the major points of a sales proposal and requests reciprocal communication. The three major points of **excellence**, **timeliness**, and **price** are themselves the three major headings in the body of the proposal.

Format

Format is how the logical order is expressed. Format governs the nature and number of "departments" within a document as well as the appearance of the proposal. Visually, proposal writing favors

◆ An "outline" form of presentation instead of an essay form
◆ The use of different fonts and sizes of type to create eye relief
◆ Flush-left text instead of tabbed indentations at the beginning of paragraphs
◆ Single-spaced paragraphs separated by double-spacing
◆ A highly-articulated structure which may divide a project into discrete, labeled parts.

In essence, every page of a technical document is a *picture*, made to appear as reader-friendly as possible.

Two types of business writing lend their formats to proposals: letters and memos. The business letter may introduce a proposal sent externally from one entity to another. Shorter proposals may themselves be written as extended business letters following the rules of block, semi-block, or modified correspondence described in technical writing manuals. Proposals sent internally may be introduced by memos, and themselves might be structured as extended memos (see Figure A.7). Working with examples of proposals you'll notice their similarities to these basic business forms, for the letter and memo structures become "shells" inside which the organization and content operate.

Figure A.7 Basic Letter and Memo Forms

```
                        ┌─────────────────────┐
                        │    Letterhead       │
                        └─────────────────────┘
┌──────────────────┐
│ Date of letter   │
└──────────────────┘
    At least two spaces down—
    More if needed to center letter

    ┌─────────────────────────────────┐
    │ Name and address of recipient   │
    └─────────────────────────────────┘
Two spaces down

    ┌─────────────────────────┐
    │ Salutation (Dear—:)     │
    └─────────────────────────┘
Two spaces down

    ┌───────────────────────────────────────────────┐
    │ Paragraph single-spaced and flush with left margin │
    └───────────────────────────────────────────────┘
One extra space between paragraphs

    ┌───────────────────────────────────────────────┐
    │ Paragraph single-spaced and flush with left margin │
    └───────────────────────────────────────────────┘
Two spaces down

    ┌─────────────────┐
    │ Closing line    │
    └─────────────────┘
Three to four spaces down to accommodate signature

    ┌──────────────────────┐
    │ Name and title, typed │
    └──────────────────────┘
Two spaces down

    ┌──────────────────┐
    │ Initials of typist │
    │ File number       │
    └──────────────────┘
Double- or single-space down

    ┌──────────────────────────────┐
    │ Enclosure: "encl" if needed  │
    └──────────────────────────────┘
Double- or single-space down

    ┌──────────────────────────────┐
    │ Copy information if necessary │
    └──────────────────────────────┘
```

Figure A.7   *(continued)*

MEMO

TO: [Recipient]
FROM: [Sender] *Initial*
DATE: [Some prefer the date line to come first]
RE: [The word "SUBJECT" may be used instead of "RE"]

The first paragraph of a memo begins two spaces down from the subject line. Paragraphs are flush with the left margin and single-spaced. This "message" paragraph should be concise.

The second paragraph begins after a blank line space. Develop details in this and any following "support" paragraphs.

Closing commentary usually requests feedback.

encl [If there are enclosures, that can be indicated here.]

This simple memorandum form becomes the foundation for several report styles, as we will see later. The basic parts of the short memo can be expanded to produce detailed ("articulated") sections of presentations.

The Structure of Formal Technical Reports.   Multipart proposals often borrow their attributes from the conventions of the formal technical report. The formal report consists of these components, usually in the following order:

1. A binder holding the report together
2. A cover sheet prepared for maximum visual impact
3. A letter or memo, called a transmittal, that addresses the audience and explains the report
4. A table of contents (followed by a table of illustrations, if important)
5. An abstract or executive summary providing the reader with a condensed overview of the contents
6. The report itself, containing an introduction, support sections, and a closing section
7. Any attachments or appendices to the report

Binding.   Such documents are often attractively bound. This enhances visual appeal, and ensures durability as long as the stiffer plastic binders are used. Cheap, flimsy binders with slide-on spines send the wrong message to your audience. You don't want your proposal coming apart in your client's hands, and you don't want your recipient to feel deemed unworthy of the few cents a better cover would cost. Plastic comb-binding works well for thicker reports, but tends to distort the shape of documents under thirty pages, making such texts look wavy. This may not be the right message to send to your client, either. An alternative for those proposals is a proprietary process in which two thin, strong plastic strips secure the pages. The strips interlock through holes punched with a special tool along the left edge of the document.

If the binder is opaque, the folder must be labeled with the title, the words "Prepared for" followed by the recipient's name, the words "Prepared by" followed by your name or your firm's name, and the date. Again: use taste in choosing a label or better yet, make one on your computer, giving it a border and using a type and font identical to that of the cover page (Figure A.8).

From the start, then, you want your report to achieve maximum impact. Close attention to the "package" itself ensures that. Whichever method you use to produce the report package, however, the result should be professional in all respects. Immediate visual appeal provides a sense of quality and attention to the client.

Cover page.   The cover sheet is the first item inside the bound package. Centered on it appear:

The Title of the Proposal

Prepared for [the Recipient]

Prepared by [You or Your Organization]

Date

Also, the cover sheet may contain an appropriate illustration to get the attention of the recipient. Illustrations that have nothing to do with the subject matter are useless, and in fact may deflect your client's attention.

Transmittal letter or memo.   Many firms place this element first, attaching it to the outside of the document. However, in the process of opening mail such items can get thrown out with the envelope. For this reason, include this letter or memo in, not on top of, the proposal. Bind it inside, right after the cover page. Remember that if the document is being sent externally, you should use a transmittal letter; if the report is an internal document, a memo carries the information.

What should the letter or memo contain? The transmittal should:

1. Address the recipient by title and surname.
2. Provide a brief summary of the background leading up to the proposal.
3. Mention the key findings of the document.
4. Explain the benefit to the reader provided by the knowledge contained in the document.
5. Indicate those not mentioned at the beginning of the letter or memo to whom copies of the report have been sent: a line at the bottom does

Figure A.8 A Proposal Cover with a Label

Proposal for Insurance and Annuity Program

Prepared for: Angela Jones, Personnel Director

Prepared by: Thomas Woodson, Broker

Date: March 23, 1999

this—just write c: (for copy) followed by names of other recipients. (You may still see cc used for this purpose, though cc technically means "carbon copy"). The purpose of doing this is ethical; people have the right to know about other recipients of apparently confidential material.

Note: If you are using a simplified letter as the transmittal, you should state the title of the report as your subject line as you would do in a memo. For reference purposes, review the examples in this book, as well as Figure A.9.

Figure A.9   A Letter and a Memo as Transmittals

SCHMENDRICK Q. GALLIWAMPUS, PRESIDENT
The Society for the Study of the Insignificant
1211 Remote Drive
Upper Finagle, MO 65551
555-555-5555
galliwampus@lowpotential.com

October 8, 2002

Branyon Banyan
Editor, Creaky Press
42 Hitchhiker's Drive
Fordham, NE 64444

Dear Mr. Banyan:

After talking with Postlequeue R. Throgmorton, the expert in identifying aberrant calls of the Turgid Loon, I have become convinced of the need for a short guidebook on that subject.

Professor Throgmorton states, and I agree, that the academic and bird-watching communities might benefit from a concise review of the many voicings possible in the call of this bird, coupled with a geographic index to the areas in which particular variants of the phrasing are likely to be heard.

Dr. Throgmorton has urged me to undertake this project and has offered his full support. I am eager to begin the work and request that you consider producing this book.

Please let me know what you think of this idea. I can be reached at the number and address above. Professor Throgmorton is sending a separate letter of support, but may also be contacted at The Center for the Incurably Avian, 345 Accipiter Rd., Eagle Ridge, IL 61533.

Sincerely,

Schmendrick Q. Galliwampus

Schmendrick Q. Galliwampus

Attachment: Proposal

Figure A.9   *(continued)*

MEMO

To: Sharon Neville, VP
From: Art Smith, Human Services *AS*
Date: April 5, 2000
Re: Day-Care Expansion

As you requested, attached is a short proposal for expanding the company day-care offerings. This proposal demonstrates that

- Expansion can be accomplished in small increments over the next five years,
- Costs can be kept close to break-even,
- There is widespread employee support for such expansion.

Please contact me at extension 4512 to discuss the issues covered in this report.

Table of contents.   This section puts parallelism to work—you don't want to mix nouns with verbs since that would deflect the reader's attention. All elements in the table of contents constitute an outline of the document using grammatically similar phrases. You want to watch the amount of detail in your table of contents, since you need to provide readers with a true picture of what your communication actually emphasizes. Absence of detail in key sections and a proliferation of detail in less-significant sections could misrepresent the scope of the proposal. You'll need to assign page numbers to the items in the table of contents so that the reader can find everything quickly (see Figure A.10).

Table of illustrations.   If there are a couple of illustrations, the best plan is to include them in a separate table below the table of contents, but not on a new sheet. Typically you'll need the description (Figure 1, Table 4), the caption of the figure, and its pagination. But if there are many illustrations, you'll need a separate page for such a table. In fact, you may have many illustrations of several types, necessitating a table which classifies them as:

◆ Figures (drawings or drawings with text)
◆ Tables (tabular arrangements of data)
◆ Plates (photographs)

Whether or not you need anything as elaborate as this classification depends on whether you have a proposal which emphasizes visual material. Figure A.10 shows a table of contents and a table of illustrations. It is no longer necessary to use periods to connect the items listed in a table with their corresponding

Figure A.10   Table of Contents and Table of Illustrations

---

page numbers, nor does one need to capitalize the first letters of all important words in the item titles. Do use an outline form, however, so that the reader can follow your organization clearly. Remember that figures are numbered decimally, with the digits to the left of the decimal denoting the chapter or section of the document and the digits to the right of the decimal representing the position of the figure in that section.

Abstract, or Executive Summary.   This MUST be a summary, normally not to exceed 250 words (see Figure A.11). You don't want to open up a large document only to find a summary of it which is itself huge. The Executive Summary must communicate to the topmost reader of your proposal the essence of that report. Employ analytical writing. Use a message, support, closure strategy; not a narrative one. And try to break up the supporting material into a bulleted list for easy digestion.

Some would qualify the above statements and suggest that if the Executive Summary runs long, a separate Abstract or Summary precede it. I would caution against doing this routinely. Why? This practice provides an abstract to summarize a summary which summarizes a report that itself has an introduction containing summary elements. Do not inflict excessive preliminary material upon busy readers. Work on your Executive Summary to make sure it reflects the gist of the proposal within 250 words. There are

Figure A.11 Example of a Short Executive Summary

<u>**Executive Summary**</u>

This document proposes a graduate degree in Multidisciplinary Studies. The concept presented derives from a "cafeteria plan" Master's degree currently offered at Schmendrick Tech (see Exhibit B), though it is adapted to our needs to include a wider scope of course offerings and extended practica. Our proposal demonstrates that such a Master's degree is feasible and marketable. This report also discusses the resources that would require improvement should we implement this degree.

exceptions, but those occur because of the formal requirements within highly-specialized fields.

Introduction.   The beginning of a semiformal proposal might explain

◆ That there *is* a problem (background)
◆ What might *fix* the problem (description of potential remedies)
◆ The specific *remedy* to be proposed (the thesis or purpose of the proposal)

People like to think in threes: remember all the folktales in which things recur in multiples of three, or the basic structure of transactional writing, with its three-part system of message, support, and closure. A tripartite introduction to a formal report may be both psychologically satisfying and memorable. For this reason, it is commonly, even instinctively, employed. Consider these patterns:

Introduction to a Problem-Solution Proposal

◆ *Overview* of problem
◆ *Statement of purpose*—the report will show how to fix the problem
◆ *Thesis* of report urging a specific solution

Introduction to a Feasibility Report

◆ *Overview* of situation
◆ *Description* of particulars designed to address the situation
◆ *Reasons* why the suggested particulars are feasible

Note the general features of these outlines of typical introductions. The introduction surveys the issue, explains the parameters of the investigation or report, and enumerates the components necessary to resolve or understand the concepts presented. That enumeration may be accomplished within a bulleted list, the items of which may be subheadings within the body of the report (see Figure A.12).

Figure A.12   A Short Introduction to a Proposal

Introduction

We request a mini-grant of $1500 twice a year to assist Schmendrick College's Community Creative Writing Project. This program would cultivate an interacting community of creative writers in a rural area, produce two creative writing collections per year, sponsor the reading of original works by local writers, post the work of such writers on the Internet, and support visits by writers to regional schools or other sites. The program would continue and enhance the present work of the Project by fostering publication and readings in the community.

The discussion section of this proposal divides as follows:

- Current challenges facing the Project
- The Project's attempts to address these challenges
- A plan for evaluating the success of the funded efforts
- Background information on institutions and individuals connected with the Project.

We hope you will consider our needs when planning your current grant disbursements.

Support section.   An understanding of claim and substantiation is crucial to the success of your support section in communicating the ideas which reinforce your message that are contained in the introduction, the abstract, and the transmittal. Generally, support in a formal report follows a structure in which assertions of fact, reinforced by values, produce statements of policy (see Figure A.13).

Figure A.13   Detail from a Support Section of a Proposal

Current Project

At present, our committee sponsors the personal awareness of wellness. Since its inception in 1990, we have reviewed, planned, and implemented poster campaigns. Each poster has a special theme: prenatal care, protection from the sun, the planning of balanced meals. Photographs of such posters appear in Appendix I. The agency also supports monthly talks by recognized speakers at the Jenkins Community Center, and produces an annual brochure, Caring for You (see enclosed issues), distributed free of charge throughout neighborhood locations such as churches, the library, and the Community Center. These brochures emphasize monitoring one's health, techniques to reduce stress, and the importance of proper diet. Response from local physicians has been completely positive (see Appendix II).

Assertions of fact are not the facts themselves, but general statements based on the specifics to be shown. For example, a topic sentence at the beginning of a paragraph may be such an assertion, supported by the specific facts which constitute the body of the paragraph. Or, as is common in transactional writing, the header itself may contain the assertion of fact which its following paragraph supports. Technical writing commonly presents the specifics in the form of bulleted lists underneath a general assertion, *or as illustrations.* Such specifics include:

◆ Statistics
◆ Case histories
◆ Definitions followed by examples
◆ Narrative or descriptive summaries
◆ Process explanations
◆ Analyses of other events
◆ Reasons derived from logical argument
◆ Charts, graphs, and tables

Values held by the audience and encouraged by the writer may help to reinforce the logic of support. For example, a proposal on the need for developing a youth recreational center probably assumes that both writer and audience desire good things for the community's children. Or, supported assertions of fact may suggest that the audience modify its value judgements in light of the new data presented. You must be sure of your audience so that you can handle the reinforcement of values properly. The hardest task in transactional writing is creating a report for an unknown audience who may be entirely skeptical about or overtly hostile to your goals.

Given the right audience reception, assertions of fact should lead to a statement of policy in a proposal with a clear goal. What do you want your audience to do? If the document's goal is the understanding of information which you present, then that is your policy. If the goal is to convince the audience to act, that action is your policy. Generally, transactional writing begins with sections devoted to analysis and presentation of facts; then it explains or reinforces the policy—the message.

Closing section. Often, the specific statement of policy appears in the closing section. If the document is informative in function, then the conclusion expresses the wish that the readers understand the information presented and apply this information to answer their needs. If the proposal is persuasive in intent, the conclusion requests either action or acceptance.

A list of any sources used (see "Documentation" in the next section) follows the conclusion. Some institutions place the list two spaces below the text of the conclusion. Others, especially for long reports, begin the list on a new page.

Back matter. A formal report always has front matter which precedes the introduction. It may contain back matter as well. Back matter, if present, typically consists of "raw" documents not integrated into the report but relevant to it. For example, a college student researching the feasibility of a day-care center on her campus might accumulate the budget proposals and the minutes of an *ad hoc* committee that had considered developing such a facility in the past. She might discuss material from the raw documents in her report, but may also wish to include the entire documents as appendices or attachments, as their presence in the report ensures credibility. She would not repaginate these documents, but would caption each with the word "Appendix" and the number I, II, or III. In the body of her document, she would refer readers to the appropriate appendix using a parenthetical reference: (see Appendix I). Back matter does not have to be limited to such material, either; surveys, marketing investigations, or brochures may find themselves at the end of the long document depending on need. Note that standard practice in business and technical writing mandates placing appendices *after* a list of sources, not before, as they would be in an academic research paper.

Documentation

You should be familiar with the principles of reference format from your English composition course; however, if you are not, consult a style handbook such as the *Prentice Hall Reference Guide to Grammar and Usage* that contains sections discussing MLA and APA requirements, as well as those of other reference formats. Or, for current MLA practice, refer to Joseph Gibaldi's *MLA Handbook for Writers of Research Papers*. Present APA format appears in the American Psychological Association's *Publication Manual*.

Documentation Systems—Citation and Reference List. You use the same documentation system throughout the report, just as you did in writing a research paper for English. If the proposal marks borrowed ideas with MLA parenthetical citation, the ending list of references should also be an MLA Works Cited page. If APA citation is employed, the reference list, in APA format, should be entitled References. Current MLA and APA practices avoid footnotes, except for explanatory material which does not fit into the text of the paper itself but is still necessary for the reader to know. Sometimes you will see a "house style" used by organizations in which the sources listed at the end are numbered as well as alphabetized; the borrowed material in the text of the report is followed by the appropriate number keying the material to the appropriate source rather than by a parenthetical citation. Though otherwise using MLA or APA format, some organizations single-space the text of each source entry in the bibliography, and double-space between entries. Follow the requirements of your institution or course for consistency; note examples in the documents throughout this book, as well as in Figure A.14, which depicts commonly-occurring "hybrid" forms.

Figure A.14   The Same Sources Modified in MLA and APA Forms

A book, an online journal article, and a magazine article appear below. The citations shown below depict "hybrid" methods in use. Check your style manuals for "pure" forms!

"MLA"

Calimari, Igor. *Incompetence on the Graduate Committee*. New York: U of Nowhere P, 2000.

Talbott, E. "Lycanthropy Revisited: Werewolves among Our Politicians." *Fang Journal* 2.2 (1999) 10 Nov. 1999 <http:// nowhere.nohow/noplace/>.

Uhlenrich, Peter. "Toenail Clippings in History." *Defeat* 23 Mar. 2000: 52.

"APA" (In Hanging Indent Style)

Calimari, I. (2000). *Incompetence on the graduate committee*. New York: University of Nowhere Press.

Talbott, E. (1999). Lycanthropy revisited: Werewolves among our politicians. *Fang Journal, 2*(2). <http:// nowhere.nohow/noplace/> (1999, November 10).

Uhlenrich, Peter. (2000, March). Toenail clippings in history. *Defeat,* 52.

What Should Get Cited?   Remember, anything not generally known to your audience, or in the realm of public knowledge, needs to be tagged with a citation—whether that is an MLA parenthetical citation of author and page number, an APA citation of author and date, or a reference number. In other words, such an idea belongs to somebody, who needs to be credited for it whether you are quoting verbatim, or paraphrasing, or summarizing to re-explain the concept.

Integrating References.   Remember too that this idea or information must be integrated into your own writing. For example, be sure to introduce a quote with a transition; a quote which stands by itself as a complete sentence looks as though it has been glued or taped onto your text. Not this example, from a choppy review of literature inside a proposal:

> "Allen measured the amount of acid rain in the Northeast" (Jones 43). [This is a "dumped quote" lacking transitional connection to the text].

Instead, this:

> One scientist, Dr. John Allen, "measured the amount of acid rain in the Northeast" (Jones 43). [An explanatory transition relates the quoted material to the text of the report].

Questions

1. Study the following research proposal in Figure A.15, in which Aimee Salmons presents a short prospectus for a forthcoming multisectioned paper on a problem in public education. Because she has planned her task carefully, this proposal's introductory summary will develop into the "message" material of her long report, the questions to be answered will evolve into the "support" sections, and the description of research will be expanded into a reference list or bibliography. Explain how this will be accomplished—in other words, state how the embryonic parts of her later paper, embedded in the proposal, will evolve:
   a. What would be added to the introductory material, and how must such material be divided into different parts?
   b. How might the parts of her hypothetical introduction reflect her mission without repeating themselves?
   c. What subordinate material might be organized under which main headings in the body of the report?
   d. What features must the final report's reference list have which are not present at this stage of the project?

2. This short proposal is modeled after a memo pattern. If it were recast into letter form, what changes would be made?

3. Can the main headers in the proposal be changed and still retain the sense of the original proposal?

4. In a longer proposal, Aimee's Introductory Summary would be expanded and might divide into an Executive Summary and an Introduction. What would those sections contain, and how might they be written?

5. In a longer proposal, supporting material will help illustrate and explain the questions to be answered. What types of material might be used?

Figure A.15   Short Proposal by Aimee Salmons

TO: Dr. Brian Holloway
FROM: Aimee Salmons
DATE: November 12, 1997
RE: Research Proposal

Introductory Summary

The purpose of my formal report will be to examine the effectiveness of current sexual education programs in schools and their communities. I also intend to focus on the significance of peer pressure, inquisitiveness, and in-experience in shaping a teenager's attitude towards sex. My thesis will read as follows:

"In order to fight such powerful enemies as peer pressure and curiosity, we have to show our children how to do what's right for them and how to know *when* it's right."

Questions to Be Answered

In this report, I would like to address the following questions:

- Is teaching abstinence as the only solution for teens effective?
- Are sex education programs sending mixed messages to students?
- Are the students more receptive to dual messages?
- Should the government be more involved in program development?
- Are existing programs being tested properly to see if they work?

Methods of Research

Since sexual education in schools is primarily a social issue, I have been relying on SIRS for most of my research. The main body of my preliminary research comes from these three articles:

Buckley, Stephen; Wilgoren, Debbi. "Young and Experienced." *Washington Post* April 24 1994: A1+. SIRS 1994 Youth, Volume Number 4, Article 67.

Daley, Daniel; Wacker, Betsy. "Sexuality and the 104th Congress: The First Hundred Days." *SIECUS Report* June/July 1995: 13-15. SIRS 1995 Sexuality, Volume Number 4, Article 52.

Roan, Shari. "Are We Teaching Too Little, Too Late?" *Los Angeles Times* July 12, 1995: E1+. SIRS 1995 Youth, Volume Number 4, Article 93.

I also conducted searches on the Internet as well as ProQuest, but did not discover any articles relating to the exact material to be discussed in my report. When I attempted to search for books in the Learning Resource Center, I could not find any current data on sex education programs or their effectiveness.

Timetable

I plan to complete my research by December 1 and finish writing my report by December 5. I then expect to have a cover sheet, letter of transmittal, table of contents, and any "back matter" completed two days before the final examination date for this class.

Request for Approval

I have now shown you what questions need to be asked about sexual education in schools, and I am also forming some hypothetical solutions to this problem with the use of my information. I hope you approve of my topic choice and allow this study to continue so I can present my information to you in the final report. Please inform me if you have any problem with this topic.

Index

A

D

G

M

N

O

P

Q

R

T

V

W